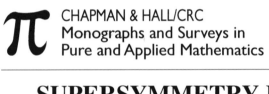

CHAPMAN & HALL/CRC
Monographs and Surveys in
Pure and Applied Mathematics   116

SUPERSYMMETRY IN

QUANTUM AND

CLASSICAL

MECHANICS

CHAPMAN & HALL/CRC
Monographs and Surveys in
Pure and Applied Mathematics   116

SUPERSYMMETRY IN

QUANTUM AND

CLASSICAL

MECHANICS

BIJAN KUMAR BAGCHI

CRC Press
Taylor & Francis Group
Boca Raton London New York

CRC Press is an imprint of the
Taylor & Francis Group, an **informa** business

A CHAPMAN & HALL BOOK

CRC Press
Taylor & Francis Group
6000 Broken Sound Parkway NW, Suite 300
Boca Raton, FL 33487-2742

First issued in paperback 2019

© 2001 by Taylor & Francis Group, LLC
CRC Press is an imprint of Taylor & Francis Group, an Informa business

No claim to original U.S. Government works

ISBN-13: 978-1-58488-197-1 (hbk)
ISBN-13: 978-0-367-39812-5 (pbk)

Library of Congress Card Number 00-059602

Library of Congress Cataloging-in-Publication Data

Bagchi, B. (Bijan Kumar)
    Supersymmetry in quantum and classical mechanics / B. Bagchi.
      p. cm.-- (Chapman & Hall/CRC monographs and surveys in pure and applied mathematics)
    Includes bibliographical references and index.
    ISBN 1-58488-197-6 (alk. paper)
    1. Supersymmetry. I. Title. II. Series.

QC174.17.S9 2000
539.7′25 --dc21   00-059602

Visit the Taylor & Francis Web site at
http://www.taylorandfrancis.com

and the CRC Press Web site at
http://www.crcpress.com

For Basabi and Minakshi

Contents

Preface

This monograph summarizes the major developments that have taken place in supersymmetric quantum and classical mechanics over the past 15 years or so. Following Witten's construction of a quantum mechanical scheme in which all the key ingredients of supersymmetry are present, supersymmetric quantum mechanics has become a discipline of research in its own right. Indeed a glance at the literature on this subject will reveal that the progress has been dramatic. The purpose of this book is to set out the basic methods of supersymmetric quantum mechanics in a manner that will give the reader a reasonable understanding of the subject and its applications. We have also tried to give an up-to-date account of the latest trends in this field. The book is written for students majoring in mathematical science and practitioners of applied mathematics and theoretical physics.

I would like to take this opportunity to thank my colleagues in the Department of Applied Mathematics, University of Calcutta and members of the faculty of PNTPM, Universite Libre de Bruxeles, especially Prof. Christiane Quesne, for their kind cooperation. Among others I am particularly grateful to Profs. Jules Beckers, Debajyoti Bhaumik, Subhas Chandra Bose, Jayprokas Chakrabarti, Mithil Ranjan Gupta, Birendranath Mandal, Rabindranath Sen, and Nandadulal Sengupta for their interest and encouragement. It also gives me great pleasure to thank Prof. Rajkumar Roychoudhury and Drs. Nathalie Debergh, Anuradha Lahiri, Samir Kumar Paul, and Prodyot Kumar Roy for fruitful collaborations. I am indebted to my students Ashish Ganguly and Sumita Mallik for diligently reading the manuscript and pointing out corrections. I also appreciate the help of Miss Tanima Bagchi, Mr. Dibyendu Bose, and Dr. Mridula

Kanoria in preparing the manuscript with utmost care. Finally, I must thank the editors at Chapman & Hall/CRC for their assistance during the preparation of the manuscript. Any suggestions for improvement of this book would be greatly appreciated.

I dedicate this book to the memory of my parents.

Bijan Kumar Bagchi

Acknowledgments

This title was initiated by the International Society for the Interaction of Mechanics and Mathematics (ISIMM). ISIMM was established in 1975 for the genuine interaction between mechanics and mathematics. New phenomena in mechanics require the development of fundamentally new mathematical ideas leading to mutual enrichment of the two disciplines. The society fosters the interests of its members, elected from countries worldwide, by a series of biannual international meetings (STAMM) and by specialist symposia held frequently in collaboration with other bodies.

CHAPTER 1

General Remarks on Supersymmetry

1.1 Background

It is about three quarters of a century now since modern quantum mechanics came into existence under the leadership of such names as Born, de Broglie, Dirac, Heisenberg, Jordan, Pauli, and Schroedinger. At its very roots the conceptual foundations of quantum theory involve notions of discreteness and uncertainty. Schroedinger and Heisenberg, respectively, gave two distinct but equivalent formulations: the configuration space approach which deals with wave functions and the phase space approach which focuses on the role of observables. Dirac noticed a connection between commutators and classical Poisson brackets and it was chiefly he who gave the commutator form of the Poisson bracket in quantum mechanics on the basis of Bohr's correspondence principle.

Quantum mechanics continues to attract the mathematicians and physicists alike who are asked to come to terms with new ideas and concepts which the tweory exposes from time to time [1-2]. Supersymmetric quantum mechanics (SUSYQM) is one such area which has received much attention of late. This is evidenced by the frequent appearances of research papers emphasizing different aspects of SUSYQM [3-9]. Indeed the boson-fermion manifestation in soluble models has considerably enriched our understanding of degeneracies

1

and symmetry properties of physical systems.

The concept of supersymmetry (SUSY) first arose in 1971 when Ramond [10] proposed a wave equation for free fermions based on the structure of the dual model for bosons. Its formal properties were found to preserve the structure of Virasoro algebra. Shortly after, Neveu and Schwarz [11] constructed a dual theory employing anticommutation rules of certain operators as well as the ones conforming to harmonic oscillator types of the conventicnal dual model for bosons. An important observation made by them was that such a scheme contained a gauge algebra larger than the Virasoro algebra of the conventional model. It needs to be pointed out that the idea of SUSY also owes its origin to the remarkable paper of Gol'fand and Likhtam [12] who wrote down tne four-dimensional Poincare super-algebra. Subsequent to these works various models embedding SUSY were proposed within a field-theoretic framework [13-14]. The most notable one was the work of Wess and Zumino [14] who defined a set of supergauge transformation in four space-time dimensions and pointed out their relevance to the Lagrangian free-field theory. It has been found that SUSY field theories prove to be the least divergent in comparison with the usual quantum field theories. From a particle physics point of view, some of the major motivations for the study of SUSY are: (i) it provides a convenient platform for unifying matter and force, (ii) it reduces the divergence of quantum gravity, and (iii) it gives an answer to the so-called "hierarchy problem" in grand unified theories.

The basic composition rules of SUSY contain both commutators and anticommutators which enable it to circumvent the powerful "no-go" theorem of Coleman and Mandula [15]. The latter states that given some basic features of S-matrix (namely that only a finite number of different particles are associated with one-particle states and that an energy gap exists between the vacuum and the one-particle state), of all the ordinary group of symmetries for the S-matrix based on a local, four-dimensional relativistic field theory, the only allowed ones are locally isomorphic to the direct product of an internal symmetry group and the Poincare group. In other words, the most general Lie algebra structure of the S-matrix contains the energy-momentum operator, the rotation operator, and a finite number of Lorentz scalar operators.

Some of the interesting features of a supersymmetric theory may be summarized as follows [16-28]:

1. Particles with different spins, namely bosons and fermions, may be grouped together in a supermultiplet. Consequently, one works in a framework based on the superspace formalism [16]. A superspace is an extension of ordinary space-time to the one with spin degrees of freedom. As noted, in a supersymmetric theory commutators as well as anticommutators appear in the algebra of symmetry generators. Such an algebra involving commutators and anticommutators is called a graded algebra.

2. Internal symmetries such as isospin or $SU(3)$ may be incorporated in the supermultiplet. Thus a nontrivial mixing between space-time and internal symmetry is allowed.

3. Composition rules possess the structure [28]

$$X_a X_b - (-)^{ab} X_b X_a = f_{ab}^c X_c$$

   where, $a, b = 0$ if X is an even generator, $a, b = 1$ if X is an odd generator, and f_{ab}^c are the structure constants. We can express X as (A, S) where the even part A generates the ordinary n-dimensional Lie algebra and the odd part S corresponds to the grading representation of A. The generalized Lie algebra with generators X has the dimension which is the sum of n and the dimension of the representation of A. The Lie algebra part of the above composition rule is of the form $T \otimes G$ where T is the space-time symmetry and G corresponds to some internal structure. Note that S belongs to a spinorial representation of a homogeneous Lorentz group which due to the spin-statistics theorem is a subgroup of T.

4. Divergences in SUSY field theories are greatly reduced. Indeed all the quadratic divergences disappear in the renormalized supersymmetric Lagrangian and the number of independent renormalization constants is kept to a minimum.

5. If SUSY is unbroken at the tree-level, it remains so to any order of $\hbar$ in perturbation theory.

In an attempt to construct a theory of SUSY that is unbroken at the tree-level but could be broken by small nonperturbative corrections, Witten [29] proposed a class of grand unified models within a field theoretic framework. Specifically, he considered models (in less than four dimensions) in which SUSY could be broken dynamically. This led to the remarkable discovery of SUSY in quantum mechanics dealing with systems less than or equal to three dimensions. Historically, however, it was Nicolai [31] who sowed the seeds of SUSY in nonrelativistic mechanics. Nicolai showed that SUSY could be formulated unambiguously for nonrelativistic spin systems by writing down a graded algebra in terms of the generators of the supersymmetric transformations. He then applied this algebra to the one-dimensional chain lattice problem. However, it must be said that his scheme did not deal explicitly with any kind of superpotential and as such connections to solvable quantum mechanical systems were not transparent.

Since spin is a well-defined concept in at least three dimensions, SUSY in one-dimensional nonrelativistic systems is concerned with mechanics describable by ordinary canonical and Grassmann variables. One might even go back to the arena of classical mechanics in the realm of which a suitable canonical method can be developed by formulating generalized Poisson brackets and then setting up a correspondence principle to derive the quantization rule. Conversely, generalized Poisson brackets can also be arrived at by taking the classical limit of the generalized Dirac bracket which is defined according to the "even" or "odd" nature of the operators.

The rest of the book is organized as follows.

In Chapter 2 we outline the basic principles of SUSYQM, starting with the harmonic oscillator problem. We try to give a fairly complete presentation of the mathematical tools associated with SUSYQM and discuss potential applications of the theory. We also include in this chapter a section on superspace formalism. In Chapter 3 we consider supersymmetric classical mechanics and study generalized classical Poisson bracket and quantization rules. In Chapter 4 we introduce the concepts of SUSY breaking and Witten index. Here we comment upon the relevance of finite temperature SUSY and analyze a regulated Witten index. We also deal with index condition and the issue of q-deformation. In Chapter 5 we provide an elaborate treatment on factorization method, shape invariance condition, and generation of solvable potentials. In Chapter 6 we deal with the radial problem and spin-orbit coupling. Chapter 7 applies SUSY to nonlinear systems and discusses a method of constructing supersymmetric KdV equation. In Chapter 8 we address parasupersymmetry and present models on it, including the one obtained from a truncated oscillator algebra. Finally, in the Appendix we broadly outline a mathematical supplement on the derivation of the form of D-dimensional Schroedinger equation.

1.2 References

[1] L.M. Ballentine, *Quantum Mechanics - A Modern Development*, World Scientific, Singapore, 1998.

[2] M. Chester, *Primer of Quantum Mechanics*, John Wiley & Sons, New York, 1987.

[3] L.E. Gendenshtein and I.V. Krive, *Sov. Phys. Usp.*, **28**, 645, 1985.

[4] A. Lahiri, P.K. Roy, and B. Bagchi, *Int. J. Mod. Phys.*, **A5**, 1383, 1990.

[5] B. Roy, P. Roy, and R. Roychoudhury, *Fortsch. Phys.*, **39**, 211, 1991.

[6] G. Levai, Lecture Notes in Physics, **427**, 127, Springer, Berlin, 1993.

[7] F. Cooper, A. Khare, and U. Sukhatme, *Phys. Rep.*, **251**, 267, 1995.

[8] G. Junker, *Supersymmetric Methods in Quantum and Statistical Physics*, Springer, Berlin, 1996.

[9] M.A. Shifman, ITEP Lectures on Particle Physics and Field Theory, **62**, 301, World Scientific, Singapore, 1999.

[10] P. Ramond, *Phys. Rev.*, **D3**, 2415, 1971.

[11] A. Neveu and J.H. Schwarz, *Nucl. Phys.*, **B31**, 86, 1971.

[12] Y.A. Gol'fand and E.P. Likhtam, *JETP Lett.*, **13**, 323, 1971.

[13] D.V. Volkov and V.P. Akulov, *Phys. Lett.*, **B46**, 109, 1973.

[14] J. Wess and B. Zumino, *Nucl. Phys.*, **B70**, 39, 1974.

[15] S. Coleman and J. Mandula, *Phys. Rev.*, **159**, 1251, 1967.

[16] A. Salam and J. Strathdee, *Fortsch. Phys.*, **26**, 57, 1976.

[17] A. Salam and J. Strathdee, *Nucl. Phys.*, **B76**, 477, 1974.

[18] V.I. Ogievetskii and L. Mezinchesku, *Sov. Phys. Usp.*, **18**, 960, 1975.

[19] P. Fayet and S. Ferrara, *Phys. Rep.*, **32C**, 250, 1977.

[20] M.S. Marinov, *Phys. Rep.*, **60C**, 1, (1980).

[21] P. Nieuwenhuizen, *Phys. Rep.*, **68C**, 189, 1981.

[22] H.P. Nilles, *Phys. Rep.*, **110C**, 1, 1984.

[23] M.F. Sohnius, *Phys. Rep.*, **128C**, 39, 1985.

[24] R. Haag, J.F. Lopuszanski, and M. Sohnius, *Nucl. Phys.*, **B88**, 257, 1975.

[25] J. Wess and J. Baggar, *Supersymmetry and Supergravity*, Princeton University Press, Princeton, NJ, 1983.

[26] P.G.O. Freund, *Introduction to Supersymmetry*, Cambridge Monographs on Mathematical Physics, Cambridge University Press, Cambridge, 1986.

[27] L. O'Raifeartaigh, Lecture Notes on Supersymmetry, *Comm. Dublin Inst. Adv. Studies*, Series A, No. 22, 1975.

[28] S. Ferrara, An introduction to supersymmetry in particle physics, *Proc. Spring School in Beyond Standard Model Lyceum Alpinum*, Zuoz, Switzerland, 135, 1982.

[29] E. Witten, *Nucl. Phys.*, **B188**, 513, 1981.

[30] E. Witten, *Nucl. Phys.*, **B202**, 253, 1982.

[31] H. Nicolai, *J. Phys. A. Math. Gen.*, **9**, 1497, 1976.

[32] H. Nicolai, *Phys. Blätter*, **47**, 387, 1991.

CHAPTER 2

Basic Principles of SUSYQM

2.1 SUSY and the Oscillator Problem

By now it is well established that SUSYQM provides an elegant description of the mathematical structure and symmetry properties of the Schroedinger equation. To appreciate the relevance of SUSY in simple nonrelativistic quantum mechanical syltems and to see how it works in these systems let us begin our discussion with the standard harmonic oscillator example. Its Hamiltonian H_B is given by

$$H_B = -\frac{\hbar^2}{2m}\frac{d^2}{dx^2} + \frac{1}{2}m\omega_B^2 x^2 \qquad (2.1)$$

where ω_B denotes the natural frequency of the oscillator and $\hbar = \frac{h}{2\pi}$, h the Planck's constant. Unless there is any scope of confusion we shall adopt the units $\hbar = m = 1$.

Associated with H_B is a set of operators b and b^+ called, respectively, the lowering (or annihilation) and raising (or creation) operators [1-6] which can be defined by $\left(p = -i\frac{d}{dx}\right)$

$$
\begin{aligned}
b &= \frac{i}{\sqrt{2\omega_B}}\left(p - i\omega_B x\right) \\
b^+ &= -\frac{i}{\sqrt{2\omega_B}}\left(p + i\omega_B x\right)
\end{aligned}
\qquad (2.2)
$$

9

Under (2.2) the Hamiltonian H_B assumes the form

$$H_B = \frac{1}{2}\omega_B \{b^+, b\} \tag{2.3}$$

where $\{b^+, b\}$ is the anti-commutator of b and b^+.

As usual the action of b and b^+ upon an eigenstate $|n >$ of harmonic oscillator is given by

$$\begin{aligned} b|n > &= \sqrt{n}|n-1 > \\ b^+|n > &= \sqrt{n+1}|n+1 > \end{aligned} \tag{2.4}$$

The associated bosonic number operator $N_B = b^+b$ obeys

$$N_B|n >= n|n > \tag{2.5}$$

with $n = n_B$.

The number states are $|n > \frac{(b^+)^n}{\sqrt{n!}}|0 > \ (n = 0, 1, 2, \ldots)$ and the lowest state, the vacuum $|0 >$, is subjected to $b|0 >= 0$.

The canonical quantum condition $[q, p] = i$ can be translated in terms of b and b^+ in the form

$$[b, b^+] = 1 \tag{2.6}$$

Along with (2.6) the following conditons also hold

$$\begin{aligned} [b, b] &= 0, \\ [b^+, b^+] &= 0 \\ [b, H_B] &= \omega_B b, \\ [b^+, H_B] &= -\omega_B b^+ \end{aligned} \tag{2.7} \tag{2.8}$$

We may utilize (2.6) to express H_B as

$$H_B = \omega_B(b^+b + \frac{1}{2}) = \omega_B\left(N_B + \frac{1}{2}\right) \tag{2.9}$$

whichfleads to the energy spectrum

$$E_B = \omega_B\left(n_B + \frac{1}{2}\right) \tag{2.10}$$

The form (2.3) implies that the Hamiltonian H_B is symmetric under the interchange of b and b^+, indicating that the associated particles obey Bose statistics.

Consider now the replacement of the operators b and b^+ by the corresponding ones of the fermionic oscillator. This will yield the fermionic Hamiltonian

$$H_F = \frac{\omega_F}{2}\left[a^+, a\right] \tag{2.11}$$

where a and a^+, identified with the lowering (or annihilation) and raising (or creation) operators of a fermionic oscillator, satisfy the conditions

$$\{a, a^+\} = 1, \tag{2.12}$$
$$\{a, a\} = 0, \quad \{a^+, a^+\} = 0 \tag{2.13}$$

We may also define in analogy with N_B a fermionic number operator $N_F = a^+ a$. However, the nilpotency conditions (2.13) restrict N_F to the eigenvalues 0 and 1 only

$$\begin{aligned}
N_F^2 &= (a^+ a)(a^+ a) \\
&= (a^+ a) \\
&= N_F \\
N_F(N_F - 1) &= 0
\end{aligned} \tag{2.14}$$

The result (2.14) is in conformity with Pauli's exclusion principle. The antisymmetric nature of H_F under the interchange of a and a^+ is suggestive that we are dealing with objects satisfying Fermi-Dirac statistics. Such objects are called fermions. As with b and b^+ in (2.2), the operators a and a^+ also admit of a plausible representation. In terms of Pauli matrices we can set

$$a = \frac{1}{2}\sigma_-, \quad a^+ = \frac{1}{2}\sigma_+ \tag{2.15}$$

where $\sigma_\pm = \sigma_1 \pm i\sigma_2$ and $[\sigma_+, \sigma_-] = 4\sigma_3$. Note that

$$\sigma_1 = \begin{pmatrix} 0 & 1 \\ 1 & 0 \end{pmatrix}, \sigma_2 = \begin{pmatrix} 0 & -i \\ i & 0 \end{pmatrix}, \sigma_3 = \begin{pmatrix} 1 & 0 \\ 0 & -1 \end{pmatrix} \tag{2.16}$$

We now use the condition (2.12) to express H_F as

$$H_F = \omega_F \left(N_F - \frac{1}{2} \right) \tag{2.17}$$

which has the spectrum

$$E_F = \omega_F \left(n_F - \frac{1}{2} \right) \tag{2.18}$$

where $n_F = 0, 1$.

For the development of SUSY it is interesting to consider [7] the composite system emerging out of the superposition of the bosonic and fermionic oscillators. The energy E of such a system, being the sum of E_B and E_F, is given by

$$E = \omega_B \left(n_B + \frac{1}{2} \right) + \omega_F \left(n_F - \frac{1}{2} \right) \tag{2.19}$$

We immediately observe from the above expression that E remains unchanged under a simultaneous destruction of one bosonic quantum $(n_B \to n_B - 1)$ and creation of one fermionic quantum $(n_F \to n_F + 1)$ or vice-versa provided the natural frequencies ω_B and ω_F are set equal. Such a symmetry is called "supersymmetry" (SUSY) and the corresponding energy spectrum reads

$$E = \omega(n_B + n_F) \tag{2.20}$$

where $\omega = \omega_B = \omega_F$. Obviously the ground state has a vanishing energy value $(n_B = n_F = 0)$ and is nondegenerate (SUSY unbroken). This zero value arises due to the cancellation between the boson and fermion contributions to the supersymmetric ground-state energy. Note that individually the ground-state energy values for the bosonic and fermionic oscillators are $\frac{\omega_B}{2}$ and $-\frac{\omega_F}{2}$, respectively, which can be seen to be nonzero quantities. However, except for the ground-state, the spectrum (2.20) is doubly degenerate.

It also follows in a rather trivial way that since the SUSY degeneracy arises because of the simultaneous destruction (or creation) of one bosonic quantum and creation (or destruction) of one fermionic quantum, the corresponding generators should behave like ba^+ (or b^+a). Indeed if we define quantities Q and Q^+ as

$$\begin{aligned} Q &= \sqrt{\omega}b \otimes a^+, \\ Q^+ &= \sqrt{\omega}b^+ \otimes a \end{aligned} \tag{2.21}$$

it is straightforward to check that the underlying supersymmetric Hamiltonian H_s can be expressed as

$$\begin{aligned}
H_s &= \omega \left(b^+ b + a^+ a \right) \\
&= \{Q, Q^+\}
\end{aligned} \qquad (2.22)$$

and it commutes with both Q and Q^+

$$\begin{aligned}
[Q, H_s] &= 0 \\
[Q^+, H_s] &= 0
\end{aligned} \qquad (2.23)$$

Further,

$$\begin{aligned}
\{Q, Q\} &= 0 \\
\{Q^+, Q^+\} &= 0
\end{aligned} \qquad (2.24)$$

Corresponding to H_s a basis in the Hilbert space composed of $H_B \otimes H_F$ is given by $\{|n> \otimes |0>_F, |n> \otimes a^+|0>_F\}$ where $n = 0, 1, 2 \ldots$ and $0 >_F$ is the fermionic vacuum.

In view of (2.23), Q and Q^+ are called supercharge operators or simply supercharges. From (2.22) - (2.24) we also see that Q, Q^+, and H_s obey among themselves an algebra involving both commutators as well as anti-commutators. As already mentioned in Chapter 1 such an algebra is referred to as a graded algebra.

It is now clear that the role of Q and Q^+ is to convert a bosonic (fermionic) state to a fermionic (bosonic) state when operated upon. This may be summarised as follows

$$\begin{aligned}
Q \, |n_B, n_F> &= \sqrt{\omega n_B} |n_B - 1, \, n_F + 1>, n_B \neq 0, n_F \neq 1 \\
Q^+ |n_B, n_F> &= \sqrt{\omega(n_B + 1)} \Big| n_B + 1, n_F - 1>, n_F \neq 0 \quad (2.25)
\end{aligned}$$

However, $Q^+|n_B, n_F >= 0$ and $Q|n_B, n_F >= 0$ for the cases ($n_B = 0, n_F = 1$) and $n_F = 0$, respectively.

To seek a physical interpretation of the SUSY Hamiltonian H_s let us use the representations (2.2) and (2.15) for the bosonic and fermionic operators. We find from (2.22)

$$H_s = \frac{1}{2} \left(p^2 + \omega^2 x^2 \right) \mathbb{1} + \frac{1}{2} \omega \sigma_3 \qquad (2.26)$$

where $\mathbb{1}$ is the (2×2) unit matrix. We see that H_s corresponds to a bosonic oscillator with an electron in the external magnetic field.

The two components of H_s in (2.26) can be projected out in a manner

$$H_+ = -\frac{1}{2}\frac{d^2}{dx^2} + \frac{1}{2}\left(\omega^2 x^2 - \omega\right) \equiv \omega b^+ b$$

$$H_- = -\frac{1}{2}\frac{d^2}{dx^2} + \frac{1}{2}\left(\omega^2 x^2 + \omega\right) \equiv \omega b b^+ \qquad (2.27a, b)$$

Equivalently one can express H_s as

$$H_s \equiv \text{diag}\,(H_-, H_+)$$
$$= \omega\left(b^+ b + \frac{1}{2}\right)\mathbb{1} + \frac{\omega}{2}\sigma_3 \qquad (2.28)$$

by making use of (2.6).

From (2.27) it is seen that H_+ and H_- are nothing but two realizations of the same harmonic oscillator Hamiltonian with constant shifts $\pm\omega$ in the energy spectrum. We also notice that $H_\pm$ are the outcomes of the products of the operators b and b^+ in direct and reverse orders, respectively, the explicit forms being induced by the representations (2.2) and (2.15). Indeed this is the essence of the factorization scheme in quantum mechanics to which we shall return in Chapter 5 to handle more complicated systems.

2.2 Superpotential and Setting Up a Supersymmetric Hamiltonian

H_+ and H_- being the partner Hamiltonians in H_s, we can easily isolate the corresponding partner potentials $V_\pm$ from (2.27). Actually these potentials may be expressed as

$$V_\pm(x) = \frac{1}{2}\left[W^2(x) \mp W'(x)\right] \qquad (2.29)$$

with $W(x) = \omega x$. We shall refer to the function $W(x)$ as the superpotential. The representations (2.29) were introduced by Witten [8] to explore the conditions under which SUSY may be spontaneously broken.

The general structure of $V_\pm(x)$ in (2.29) is indicative of the possibility that we can replace the coordinate x in (2.27) by an arbitrary function $W(x)$. Indeed the forms (2.29) of $V_\pm$ reside in the following general expression of the supersymmetric Hamiltonian

$$H_s = \frac{1}{2}\left(p^2 + W^2\right)\mathbb{1} + \frac{1}{2}\sigma_3 W' \tag{2.30}$$

$W(x)$ is normally taken to be a real, continuously differentiable function in $\Re$. However, should we run into a singular $W(x)$, the necessity of imposing additional conditions on the wave functions in the given space becomes important [10].

Corresponding to H_s, the associated supercharges can be written in analogy with (2.21) as

$$\begin{aligned} Q &= \frac{1}{\sqrt{2}}\begin{pmatrix} 0 & W + ip \\ 0 & 0 \end{pmatrix} \\ Q^+ &= \frac{1}{\sqrt{2}}\begin{pmatrix} 0 & 0 \\ W - ip & 0 \end{pmatrix} \end{aligned} \tag{2.31}$$

As in (2.22), here too Q and Q^+ may be combined to obtain

$$H_s = \{Q, Q^+\} \tag{2.32}$$

Furthermore, H_s commutes with both Q and Q^+

$$\begin{aligned}[] [Q, H_s] &= 0 \\ [Q^+, H_s] &= 0 \end{aligned} \tag{2.33}$$

Relations (2.30) - (2.33) provide a general nonrelativistic basis from which it follows that H_s satisfies all the criterion of a formal supersymmetric Hamiltonian. It is obvious that these relations allow us to touch upon a wide variety of physical systems [12-53] including approximate formulations [54-63].

In the presence of the superpotential $W(x)$, the bosonic operators b and b^+ go over to more generalized forms, namely

$$\begin{aligned} \sqrt{2\omega}\, b \to A &= W(x) + \frac{d}{dx} \\ \sqrt{2\omega}\, b^+ \to A^+ &= W(x) - \frac{d}{dx} \end{aligned} \tag{2.34}$$

In terms of A and A^+ the Hamiltonian H_s reads

$$2H_s = \frac{1}{2}\{A, A^+\}\mathbb{1} + \frac{1}{2}\sigma_3 [A, A^+] \tag{2.35}$$

Expressed in a matrix structure H_s is diagonal

$$H_s \equiv \text{diag } (H_-, H_+)$$
$$= \frac{1}{2} \text{ diag } (AA^+, A^+A) \tag{2.36}$$

Note that H_s as in (2.30) is just a manifestation of (2.34). In the literature it is customery to refer to H_+ and H_- as "bosonic" and "fermionic" hands of H_s, respectively.

The components $H_\pm$, however, are deceptively nonlinear since any one of them, say H_-, can always be brought to a linear form by the transformation $W = u'/u$. Thus for a suitable u, $W(x)$ may be determined which in turn sheds light on the structure of the other component.

It is worth noting that both $H_\pm$ may be handled together by taking recourse to the change of variables $W = gu'/u$ where, g, which may be positive or negative, is an arbitrary parameter. We see that $H_\pm$ acquire the forms

$$2H_\pm = -\frac{d^2}{dx^2} + \left(g^2 \pm g\right) \left(\frac{u'}{u}\right)^2 \mp g \left(\frac{u''}{u}\right) \tag{2.37}$$

It is clear that the parameter g effects an interchange between the "bosonic" and "fermionic" sectors : $g \to -g, H_+ \leftrightarrow H_-$. To show how this procedure works in practice we take for illustration [64] the superpotential conforming to supersymmetric Liouville system [24] described by the superpotential $W(x) = \frac{\sqrt{2}g}{a} \exp\left(\frac{ax}{2}\right)$, g and a are parameters. Then u is given by $u(x) = \exp\left[2\sqrt{2} \left\{\exp\left(\frac{ax}{2}\right)\right\} / a^2\right]$. The Hamiltonian H_+ satisfies $\left(-\frac{d^2}{dx^2} + W^2 - W'\right)\psi_+ = 2E_+\psi_+$. Transforming $y = \frac{4\sqrt{2}}{a^2} g \exp\left(\frac{ax}{2}\right)$, the Schroedinger equation for H_+ becomes

$$\frac{d^2}{dy^2}\psi_+ + \frac{1}{y}\frac{d}{dy}\psi_+ + \left(\frac{1}{2g} - \frac{1}{4}\right)\psi_+ + \frac{8E_+}{a^2y^2}\psi_+ = 0 \tag{2.38}$$

The Schroedinger equation for H_- can be at once ascertained from (2.38) by replacing $g \to -g$ which means transforming $y \to -y$. The relevant eigenfunctions turn out to be given by confluent hypogeo-metric function.

The construction of the SUSYQM scheme presented in (2.30) - (2.33) remains incomplete until we have made a connection to the Schroedinger Hamiltonian H. This is what we'll do now.

Pursuing the analogy with the harmonic oscillator problem, specifically (2.27a), we adopt for V the form $V = \frac{1}{2}\left(W^2 - W'\right) + \lambda$ in Wwhich the constant λ can be adjusted to coincide with the ground-state energy E_0 oh H_+. In other words we write

$$V(x) - E_0 = \frac{1}{2}\left(W^2 - W'\right) \qquad (2.39)$$

indicating that V and V_+ can differ only by the amount of the ground-state energy value E_0 of H.

If $W_0(x)$ is a particular solution, the general solution of (2.39) is given by

$$W(x) = W_0(x) + \frac{\exp\left[2\int^x W_0(\tau)d\tau\right]}{\beta - \int^x \exp\left[2\int^y W_0(\tau)d\tau\right]dy}, \quad \beta \in \mathbb{R} \qquad (2.40)$$

On the other hand, the Schroedinger equation

$$\left[-\frac{1}{2}\frac{d^2}{dx^2} + V(x) - E_0\right]\psi_0 = 0 \qquad (2.41)$$

subject to (2.39) has the solution

$$\psi_0(x) = A\exp\left[-\int^x W(\tau)d\tau\right] + B\exp\left[-\int^x W(\tau)d\tau\right]$$
$$\int^x \exp\left[2\int^y W(\tau)d\tau\right]dy \qquad (2.42)$$

where $A, B, \in \mathbb{R}$ and assuming $\psi(x) \in L^2(-\infty, \infty)$. If (2.40) is substituted in (2.42), the wave function is the same [65] whether a particular $W_0(x)$ or a general solution to (2.39) is used in (2.42).

In $N = 2$ SUSYQM, in place of the supercharges Q and Q^+ defined in (2.31), we can also reformulate the algebra (2.32) - (2.35) by introducing a set of hermitean operators Q_1 and Q_2 being expressed as

$$Q = (Q_1 + iQ_2)/2, \quad Q^+ = (Q_1 - iQ_2)/2 \qquad (2.43)$$

While (2.32) is converted to $H_s = Q_1^2 = Q_2^2$ that is

$$\{Q_i, Q_j\} = 2\delta_{ij}H_s \qquad (2.44)$$

(2.33) becomes

$$[Q_i, H_s] = 0, \quad i = 1, 2 \tag{2.45}$$

In terms of the superpotential $W(x), Q_1$ and Q_2 read

$$
\begin{aligned}
Q_1 &= \frac{1}{\sqrt{22}} \left(\sigma_1 W - \sigma_2 \frac{p}{\sqrt{m}} \right) \\
Q_2 &= \frac{1}{\sqrt{22}} \left(\sigma_1 \frac{p}{\sqrt{m}} + \sigma_2 W \right)
\end{aligned} \tag{2.46}
$$

On account of (2.45), Q_1 and Q_2 are constants of motion: $\dot{Q}_1 = 0$ and $\dot{Q}_2 = 0$.

From (2.44) we learn that the energy of an arbitrary state is strictly nonnegative. This is because [66]

$$
\begin{aligned}
E_\psi &= \, < \psi | H_s | \psi > \\
&= \, < \psi | Q_1^\dagger Q_1 | \psi > \\
&= \, < \phi | \phi > \geq 0
\end{aligned} \tag{2.47}
$$

where $|\phi> = Q_1|\psi>$, and we have used in the second step the representation (2.44) of H_s.

For an exact SUSY

$$
\begin{aligned}
Q_1|0> &= 0 \\
Q_2|0> &= 0
\end{aligned} \tag{2.48}
$$

So $|\phi> \neq 0$ would mean existence of degenerate vacuum states $|0>'$ and $|0>$ related by a supercharge signalling a spontaneous symmetry breaking.

It is to be stressed that the vanishing vacuum energy is a typical feature of unbroken SUSY models. For the harmonic oscillator whose Hamiltonian is given by (2.3) we can say that H_B remains invariant under the interchange of the operators b and b^+. However, the same does not hold for its vacuum which satisfies $b|0>$. In the case of unbroken SUSY both the Hamiltonian H_s and the vacuum are invariant with respect to the interchange $Q \leftrightarrow Q^+$.

2.3 Physical Interpretation of H_s

As for the supersymmetric Hamiltonian in the oscillator case here also we may wish to seek [66, 36] a physical interpretation of (2.30).

To this end let us restore the mass parameter m in H_s which then reads

$$H_s = \frac{1}{2} \left(\frac{p^2}{m} + W^2 \right) \mathbb{1} + \frac{1}{2} \sigma_3 \frac{W'}{\sqrt{m}} \tag{2.49}$$

Comparing with the Schroedinger Hamiltonian for hhe electron (mass m and charge $-e$) subjected to an external magnetic field namely

$$\mathcal{H} = \frac{1}{2} \left(\frac{p^2}{m} + \frac{e^2}{m} \vec{A}^2 \right) + \frac{ie}{2m} div \vec{A} - \frac{e}{m} \vec{A}.\vec{p} + \frac{|e|}{2m} \vec{\sigma}.\vec{B} \tag{2.50}$$

where $\vec{A} = \frac{1}{2} \vec{B} \times \vec{r}$ is the vector potential, we find that (2.50) goes over to (2.49) for the specific case when $\vec{A} = \left(0. \frac{\sqrt{m}}{2|e|} W, 0 \right)$. The point to observe is the importance of the electron magnetic moment term in (2.50) without which it is not reducible to (2.49). We thus see that a simple problem of an electron in the external magnetic field exhibies SUSY.

Let us dwell on the Hamiltonian $\mathcal{H}$ a little more. If we assume the magnetic field $\vec{B}$ to be constant and parallel to the Z axis so that $\vec{B} = B\hat{k}$, it follows that

$$\vec{A}.\vec{p} = \frac{1}{2} BL_z$$

$$4\vec{A}^2 = r^2 B^2 - \left(\vec{r}.\vec{B} \right)^2$$

$$= \left(x^2 + y^2 \right) B^2 \tag{2.51}$$

As a result $\mathcal{H}$ becomes

$$\mathcal{H} = \frac{1}{2m} \left[p_z^2 + (p_x^2 + p_y^2) \right] + \frac{1}{2} m\omega^2 \left(x^2 + y^2 \right) - \omega \left(L_z - \sigma_3 \right) \tag{2.50a}$$

Apart from a free motion in the z direction, $\mathcal{H}$ describes two harmonic oscillators in the xy-plane and also involves a coupling to the orbital and spin moments. In (2.50a) ω is the Larmor frequency: $\omega = \frac{eB}{2m}$ and $\vec{S} = \frac{1}{2} \vec{\sigma}$.

In the standard approach of quantization of oscillators the coupling terms look like

$$\omega \left(L_z - \sigma_3 \right) = -i\omega \left(b_x^+ b_y - b_y^+ b_x \right) + \omega \sigma_3 \tag{2.52}$$

However, setting

$$B^+ = \frac{1}{\sqrt{2}} \left(b_x^+ + i b_y^+ \right)$$

$$B = \frac{1}{\sqrt{2}} \left(b_x - i b_y \right) \tag{2.53}$$

we may diagonilize (2.50a) to obtain $\frac{1}{2} \left(H - \frac{p_z^2}{2m} \right) = \omega \left(B^+ B + \frac{1}{2} \right)$ $\mathbb{1} + \frac{\omega}{2}\sigma_3$ which is a look-alike of (2.28). Summarizing, the two-dimensional Pauli equation (2.50) gives a simple illustration of how SUSY can be realized in physical systems.

2.4   Properties of the Partner Hamiltonians

As interesting property of the supersymmetric Hamiltonian H_s is that the partner components H_+ and H_- are almost isopectral. Indeed if we set

$$H_+ \psi_n^+ = E_n^+ \psi_n^+ \tag{2.54}$$

it is a simple exercise to work out

$$
\begin{aligned}
H_- \left(A \psi_n^+ \right) &= \frac{1}{2} A A^+ \left(A \psi_n^+ \right) \\
&= A \left(\frac{1}{2} A^+ A \psi_n^+ \right) \\
&= E_n^+ \left(A \psi_n^+ \right)
\end{aligned}
\tag{2.55}
$$

This clearly shows E_n^+ to be the energy spectra of H_- also. However, $A\psi_0^+$ is trivially zero since ψ_0^+ being the ground-state solution of H_+ satisfies

$$-(\psi_0^+)'' + (W^2 - W')\psi_0^+ = 0 \tag{2.56a}$$

and so is constrained to be of the form

$$\psi_0^+ = C \exp\left(-\int^x W(y)dy \right) \tag{2.56b}$$

C is a constant.

We conclude that the spectra of H_+ and H_- are identical except for the ground state ($n = 0$) which is nondegenerate and, in the present setup, is with the H_+ component of H_s. This is the case of unbroken SUSY (nondegenerate vacuum). However, if SUSY were to be broken (spontaneously) then H_+ along with H_- can not posses any normalizable ground-state wave function and the spectra of H_+ and H_- would be similar. In other words the nondegeneracy of the ground-state will be lost.

For square-integrability of ψ_0 in one-dimension we may require from (2.56) that $\int W(y)dy \to \infty$ as $|x| \to \infty$. One way to realize this condition is to have $W(x)$ differing in sign at $x \to \pm\infty$. In other words, $W(x)$ should be an odd function. As an example we may keep in mind the case $W(x) = \omega x$. On the other hand, if $W(x)$ is an even function, that is it keeps the same sign at $x \to \pm\infty$, the square-integrability condition cannot be fulfilled. A typical example is $W(x) = x^2$.

From (2.54) and (2.55) we also see for the following general eigenvalue problems of $H_\pm$

$$H_+\psi_{n+1}^{(+)} = E_{n+1}^{(+)}\psi_{n+1}^{(+)}$$

$$H_-\psi_n^{(-)} = E_n^{(-)}\psi_n^{(-)} \qquad (2.57a,b)$$

that if $A\psi_0^+ = 0$ holds for a normalizable eigenstate ψ_0^+ of H_+, then since $H_+\psi_0^+ \equiv \frac{1}{2}A^+A\psi_0^+ = 0$, it follows that such a normalizable eigenstate is also the ground-state of H_+ with the eigenvalue $E_0^+ = 0$. Of course, because of the arguments presented earlier, H_- does not possess any normalHzed eigenstate with zero-energy value.

To inquire how the spectra and wave functions of H_+ and H_- are related we use the decompositions (2.36) to infer from (2.57) the eigenvalue equations

$$H_+\left(A^+\psi_n^-\right) = \frac{1}{2}A^+A\left(A^+\psi_n^-\right) = A^+H_-\psi_n^- = E_n^-\left(A^+\psi_n^-\right) \quad (2.58a)$$

$$H_-\left(A\psi_n^+\right) = \frac{1}{2}AA^+\left(A\psi_n^+\right) = AH_+\psi_n^+ = E_n^+\left(A\psi_n^+\right) \qquad (2.58b)$$

It is now transparent that the spectra and wave functions of H_+ and H_- are related a la [52]

$$E_n^- = E_{n+1}^+, \quad n = 0,1,2,\ldots; \quad E_0^+ = 0 \qquad (2.59a)$$

$$\psi_n^- = \left(2E_{n+1}^+\right)^{-\frac{1}{2}}A\psi_{n+1}^+ \qquad (2.59b)$$

$$\psi_{n+1}^+ = \left(2E_n^-\right)^{-\frac{1}{2}}A^+\psi_n^- \qquad (2.59c)$$

We now turn to some applications of the results obtained so far.

2.5   Applications

(a) SUSY and the Dirac equation

One of the important aspects of SUSY is that it appears naturally in the first quantized massless Dirac operator in even dimensions. To examine this feature [47, 54-74] we consider the Dirac equation in $(1+2)$ dimensions with minimal electromagnetic coupling

$$\left(i\gamma^{\mu}\mathcal{D}_{\mu} - m\right)\psi = 0 \qquad (2.60)$$

where $\mathcal{D}_{\mu} = D_{\mu} + iqA_{\mu}$ with $q = -|e|$. The γ matrices may be realized in terms of the Pauli matrices since in $(1+2)$ dimensions (2.60) can be expressed in a 2×2 matrix form: $\gamma_{0} = \sigma_{3}, \gamma_{1} = i\sigma_{1}$ and $\gamma_{2} = i\sigma_{2}$. Introducing covariant derivatives

$$\begin{aligned}
D_{1} &= \frac{\partial}{\partial x} - ieA_{1}, \\
D_{2} &= \frac{\partial}{\partial y} - ieA_{2}
\end{aligned} \qquad (2.61)$$

Then (2.60) translates, in the massless case, to

$$-\left(\sigma_{1}D_{1} + \sigma_{2}D_{2}\right)\psi = \sigma_{3}E\psi \qquad (2.60a)$$

The above equation is also representative of

$$\begin{pmatrix} 0 & A \\ A^{+} & 0 \end{pmatrix}\psi = -\sigma_{3}E\psi \qquad (2.62)$$

where $A = D_{1} - iD_{2}$ and $A^{+} = D_{1} + iD_{2}$. From (2.35) and (2.36) we therefore conclude

$$2H_{s}\psi = E^{2}\psi \qquad (2.63)$$

The supersymmetric Hamiltonian thus gives the same eimenfunction and square of the energy of the original massless equation. This also makes clear the original curiosity [75] of SUSY which was to consider the "square root" of the Dirac operator in much the same manner as the "square-root" of the Klein-Gordon operator was utilized to arrive at the Dirac equation. In the case of a massive fermion the eigenvalue in (2.63) gets replaced as $E^{2} \to E^{2} - m^{2}$.

In connection with the relation between chiral anomaly and fermionic zero-modes, Jackiw [68] observed some years ago that the Dirac Hamiltonian for (2.60), namely

$$H = \vec{\alpha}.\left(\vec{p} + e\vec{A}\right) \qquad (2.64)$$

where $\vec{\alpha} = (-\sigma^2, \sigma^1)$, displays a conjugation-symmetric spectrum with zero-modes under certain conditions for the background field. The symmetry, however, is broken by the appearance of a mass term. Actually, in a uniform magnetic field the square of H coincides with the Pauli Hamiltonian. As already noted by us the latter exhibits SUSY which when exact possesses a zero-value nondegenerate vacuum.

Hughes, Kostelecky, and Nieto [69] have studied SUSY of massless Dirac operator in some detail by focussing upon the role of Foldy-Wouthusen (FW) transformations and have demonstrated the relevance of SUSY in the first-order Dirac equation. To bring out Dirac-FW equivalence let us follow the approach of Beckers and Debergh [71]. These authors have pointed out that since SUSYQM is characterized by the algebra (2.32) and (2.33) involving odd super-charges, it is logical to represent the Dirac Hamiltonian as a sum of odd and even parts

$$H_D = Q_1 + \beta m \qquad (2.65)$$

where Q_1 is odd and the mass term being even has an attached multiplicative coefficient β that anticommutes with Q_1

$$\{Q_1, \beta\} = 0 \qquad (2.66)$$

Squaring (2.65) at once yields

$$\begin{aligned} H_D^2 &= Q_1^2 + m^2 \\ &= H_s + m^2 \end{aligned} \qquad (2.67)$$

from (2.44).

We an interpret (2.67) from the point of view of FW transformation which works as

$$\begin{aligned} H_{FW} &= U\,H_D U^{-1} \\ &= \beta (H_s + m^2)^{1/2} \end{aligned} \qquad (2.68)$$

implying that the square of H_{FW} is just proportional to the right-hand side of (2.67). Note that U, which is unitary, is given by

$$S = S^+ : U = \exp(iS)$$
$$S = -\frac{i}{2}\beta Q_1 K^{-1}\theta$$
$$\tan\theta = \frac{K}{m}$$
$$[\theta, \beta] = 0$$
$$\{H_D, S\} = 0 \tag{2.69}$$

where K is even and stands for $\sqrt{H_s}$ with the positive sign. We can also write

$$U = \frac{E + \beta Q_1 + m}{[2E(E + m)]^{1/2}} \tag{2.70}$$

with $E = (H_s + m^2)^{1/2}$.

The SUSY of he massless Dirac operator links directly to two very important fields in quantum theory, namely index theorems and anomalies. Indeed it is just the asymmetry of the Dirac ground state that leads to these phenomena.

(b) SUSY and the construction of reflectionless potentials

In quantum mechanics it is well known that symmetric, reflectionless potentials provide good approximations to confinement and their constructions have always been welcome [48,76-78]. In the following we demonstrate [76-86] how the ideas of SUSYQM can be exploited to derive the forms of such potentials.

Of the two potentials $V_\pm$, let us impose upon V_- the criterion that it possesses no bound state. So we take it to be a constant $\frac{1}{2}\chi^2$

$$V_- \equiv \frac{1}{2}\left(W^2 + W'\right) = \frac{1}{2}\chi^2 > 0 \tag{2.71}$$

Equation (2.71) can be linearized by a substitution $W = g'/g$ which converts it to the form

$$\frac{g''}{g} = \chi^2 \tag{2.72}$$

The solution of (2.72) can be used to determine $W(x)$ as

$$W(x) = \chi \tanh \chi(x - x_0) \tag{2.73}$$

Knowing $W(x), V_+$ can be ascertained to be

$$\begin{aligned} V_+ &\equiv \frac{1}{2}\left(W^2 - W'\right) \\ &= \frac{1}{2}\chi^2\left[1 - 2\text{sech}^2\chi\,(x - x_0)\right] \end{aligned} \qquad (2.74)$$

One can check that H_+ possesses a zero-energy bound state wave function given by

$$\psi_0 \sim \frac{1}{g} \sim \text{sech}\chi(x - x_0) \qquad (2.75)$$

becausc

$$H_+\psi_0 = \frac{1}{2}\left[-\psi_0'' + \left(W^2 - W'\right)\psi_0\right] = 0 \qquad (2.76)$$

corresponding to the solution given in (2.73).

All this can be generalized by rewriting the previous steps as follows. We search for a potential V_1 that satisfies the Schroedinger equation

$$\left(-\frac{1}{2}\frac{d^2}{dx^2} + V_1\right)\psi_1 = -\chi_1^2\psi_1 \qquad (2.77)$$

with $V_1 + \chi_1^2$ signifying a zero-energy bound state. The relation (2.74) is re-expressed as

$$W_1^2 - W_1' = V_! + \chi_1^2 \qquad (2.78)$$

with χ_1 obtained from $W_1^2 + W_1' = \chi_1^2$ and V_1 identified with $-2\chi_1^2\text{sech}^2$ $\chi_1(x - x_0)$. Note that $V_1(\pm\infty) = 0$. Further requring $Vf1$ to be symmetric means that W_1 must be an odd function.

For an arbitrary n levels, we look for a chain of connections

$$W_n^2 + W_n' = V_{n-1} + \chi_n^2 \qquad (2.79)$$

with V_{n-1} assumed to be known (notethat $V_0 = 0$). Then V_n is obtained from

$$W_n^2 - W_n' = V_n + \chi_n^2 \qquad (2.80)$$

where $W_n(0)$ is taken to be vanishing.

Linearization of (2.79) is accomplished by the substition $W_n = g_n'/g_n$ yielding

$$-g_n'' + V_{n-1}g_n = -\chi_n^2 g_n \qquad (2.81)$$

As with (2.75), has also $U_n = \frac{1}{g_n}$ satisfies

$$- U_n'' + (W_n^2 - W_n')u_n = -u_n'' + (V_n + \chi_n^2)U_n = 0 \qquad (2.82)$$

That is

$$- U_n'' + V_n U_n = -\chi_n^2 U_n \qquad (2.83)$$

which may be looked upon as a generalization of (2.76) to n-levels. In this way one arrves at a form of the Schroedinger equation which has n distinct eigenvalues. Evidently $V_1 = -2\chi_1^2 \mathrm{sech}^2 \, \chi_1(x - x_0)$ is reflectionless.

In the study of nonlinear systems, V_1 can be regarded as an instantaneous frozen one-soliton solution of the KdV equation $u_t = -u_{xxx} + 6uu_x$. The n-soliton solution of the KdV, similarly, also emerges [79-85] as families of reflectionless potentials. It may be remarked that if we solve (2.81) and use $g_n(x) = g_n(-x)$ then we uniquely determine $W_n(x)$. For a further discussion of the construction of reflectionless potentials supporting a prescribed spectra of bound states we refer to the work of Schonfeld et al. [85].

(c) SUSY and derivation of a hierarchy of Hamiltonians

The ideas of SUSYQM can also be used to derive a chain of Hamiltonians having the properties that the adjacent members of the hierarchy are SUSY partners. To look into this we first note that an important consequence of the representations (2.29) is that the partner potentials $V_\pm$ are related through

$$V_+(x) = V_-(x) + \frac{d^2}{dx^2} \log \psi_0^+(x) \qquad (2.84)$$

where we have used (2.56). The above equation implies that once the properties of $V_-(x)$ are given, those of $V_+(x)$ become immediately known. Actually in our discussion of reflectionless potentials we exploited this feature.

We now proceed to generate a sequence of Hamiltonians employing the preceding results of SUSY. Sukumar [29] pointed out that if a certain one-dimensional Hamiltonian having a potential $V_1(x)$ allows for M bound states and has the ground-state eigenvalue and eigenfunction as $E_0^{(i)}$ and $\psi_0^{(i)}$, respectively, one can express this Hamiltonian in a similar form as H_+

$$
\begin{aligned}
H_1 &= -\frac{1}{2}\frac{d^2}{dx^2} + V_1(x) \\
&= \frac{1}{2}A_1^+ A_1 + E_0^{(i)}
\end{aligned}
\tag{2.85}
$$

where A_1 and A_1^+ are defined in terms of $\psi_0^{(1)}$. Using (2.34) and (2.56) A_1 and A_1^+ can be expressed as

$$
\begin{aligned}
A_1 &= \frac{d}{dx} - \left[\psi_0^{(1)}\right]' / \psi_0^{(1)} \\
A_1^+ &= -\frac{d}{dx} - \left[\psi_0^{(1)}\right]' / \psi_0^{(1)}
\end{aligned}
\tag{2.86}
$$

where a prime denotes a derivative with respect to x.

The supersymmetric partner to H_1 is obtained simply by interchanging the operators A_1 and A_1^+

$$
H_2 = -\frac{1}{2}\frac{d^2}{dx^2} + V_2(x) = \frac{1}{2}A_1 A_1^+ + E_0^{(1)}
\tag{2.87}
$$

where the correlation between V_1 and V_2 is provided by (2.84)

$$
V_2(x) = V_1(x) - \frac{d^2}{dx^2}ln\psi_0^{(1)} = V_1(x) + \left[A_1, A_1^+\right]
\tag{2.88}
$$

From (2.59) we can relate the eigenvalues and eigenfunctions of H_1 and H_2 as

$$
\begin{aligned}
E_{n+1}^{(1)} &= E_n^{(2)} \\
\psi_n^{(2)} &= \left[2E_{n+1}^{(1)} - 2E_0^{(1)}\right]^{-1/2} A_1 \psi_{n+1}^{(1)}
\end{aligned}
\tag{2.89}
$$

To generate a hierarchy of Hamiltonians we put H_2 in place of H_1 and carry out a similar set of operations as we have just now done. It turns out that H_2 can be represented as

$$
H_2 = -\frac{1}{2}\frac{d^2}{dx^2} + V_2(x) = \frac{1}{2}A_2^+ A_2 + E_0^{(2)}
\tag{2.90}
$$

with

$$
\begin{aligned}
A_2 &= \frac{d}{dx} - \left[\psi_0^{(2)}\right]' / \psi_0^{(2)} \\
A_2^+ &= -\frac{d}{dx} - \left[\psi_0^{(2)}\right]' / \psi_0^{(2)}
\end{aligned}
\tag{2.91}
$$

H_2 induces for itself a supersymmetric partner H_3 which can be obtained by reversing the order of the operators A_2^+ and A_2. In this way we run into H_4 and build up a sequence of Hamiltonian $H_4, H_5, \ldots$ etc. A typical H_n in this family reads

$$H_n = -\frac{1}{2}\frac{d^2}{dx^2} + V_n(x) = \frac{1}{2}A_n^+ A_n + E_0^{(n)} = A_{n-1}A_{n-1}^+ + E_0^{(n-1)} \quad (2.92)$$

with

$$A_n = \frac{d}{dx} - \left[\psi_0^{(n)}\right]' / \psi_0^{(n)}$$

$$A_n^+ = -\frac{d}{dx} - \left[\psi_0^{(n)}\right]' / \psi_0^{(n)} \quad (2.93)$$

and having the potential V_n

$$V_n(x) = V_{n-1}(x) - \frac{d^2}{dx^2} \ln \psi_0^{(n-1)}, \quad n = 2, 3, \ldots M \quad (2.94)$$

Further, the eigenvalues and eigenfunctions of H_n are given by

$$E_m^{(n)} = E_{m+1}^{(n-1)} = \ldots = E_{(m+n-1)}^{(1)},$$

$$m = 0, 1, 2, \ldots M - n, \quad n = 2, 3, \ldots M \quad (2.95)$$

$$\psi_m^{(n)} = \left\{\left[2E_{m+n-1}^{(1)} - 2E_{n-2}^{(1)}\right]\left[2E_{m+n-1}^{(1)} - 2E_{n-3}^{(1)}\right] \ldots\right.$$

$$\left.\left[2E_{m+n-1}^{(1)} - 2E_0^{(1)}\right]\right\}^{-\frac{1}{2}} \times A_{n-1}A_{n-2} \ldots A_1 \psi_1^{(n+m-1)} \quad (2.96)$$

The following two illustrations will make clear the generation of Hamiltonian hierarchy.

(1) Harmonic oscillator

Take $V_1 = \frac{1}{2}\omega^2 x^2$. The ground-state wave function is known to be $\psi_1^{(0)} \sim e^{-\omega x^2/2}$. It follows from (2.84) that $V_2(x) = V_1(x) + \omega$, $V_3(x) = V_2(x) + \omega = V_1(x) + 2\omega$ etc. leading to $V_k(x) = V_1(x) + (k - 1)\omega$. This amounts to a shifting of the potential in units of ω.

(2) Particle in a box problem

Here the relevant potential is given by

$$V_1 = 0 \quad |x| < a$$

$$= \infty \quad |x| = a$$

The energy spectrum and ground-state wave function are well known

$$
\begin{aligned}
E_m^{(1)} &= \frac{\pi^2}{8a^2}(m+1)^2, \ m = 0,1,2\ldots \\
\psi_0^{(1)} &= A\cos\frac{\pi x}{2a}
\end{aligned}
$$

where A is a constant. From (2.89) we find for V_n the result

$$
\begin{aligned}
V_n(x) &= V_1(x) + \frac{\pi}{8a}n(n-1)\sec^2\frac{\pi x}{2a} \quad n = 1,2,3\ldots \\
E_m^{(n)} &= E_{m+n-1}^{(1)} = \frac{\pi^2}{8a^2}(n+m)^2 \quad m = 0,1,2\ldots
\end{aligned}
$$

We thus see that the "particle in the box" problem generates a series of $\sec^2\frac{\pi x}{2a}$ potentials. The latter is a well-studied potential in quantum mechanics and represents an exactly solvable system.

(d) SUSY and the Fokker-Planck equation

As another example of SUSY in physical systems let us examine its subtle role [18] on the evaluation of the small eigenvalue associated with the "approach to equilibrium" problem in a bistable system. For a dissipative system under a random force $F(t)$ we have the Langevin equation

$$
\dot{x} = -\frac{\partial U}{\partial x} + F(t) \tag{2.97}
$$

where U is an arbitrary function of x and $F(t)$ depicts the noise term. Assuming $F(t)$ to have the "white-noise" correlation $(\beta = \frac{1}{T})$

$$
\langle F(t)\rangle = 0, \ \langle F(t)F(t')\rangle = 2\beta\delta(t-t') \tag{2.98}
$$

the probability of finding $F(t)$ becomes Gaussian

$$
P\left[F(t)\right] = A\exp\left[-\frac{1}{2\beta}\int F^2(t)dt\right] \tag{2.99}
$$

where $A^{-1} = \int D[F]e^{-\frac{1}{2\beta}\int F^2(t)dt}$.

The Fokker-Planck eqution for the probability distribution P is given by [87]

$$
\frac{\partial P}{\partial t} = \frac{\partial}{\partial x}\left(P\frac{\partial U}{\partial x}\right) + \beta\frac{\partial^2 P}{\partial x^2} \tag{2.100}
$$

Equation (2.100) can be converted to the Schroedinger form by enforcing the transformation

$$P = \sqrt{P_{eq}}\psi$$
$$= e^{-U/2\beta}\psi \tag{2.101}$$

where $P_{eq} = P_0 e^{-U/\beta}$ and the normalization condition $\int P_{eq}(x)dx = 1$ fixes $P_0 = 1$. Note that by setting $\frac{\partial P}{\partial t} = 0$ we get the equilibrium distribution. Setting $\psi = e^{-\lambda t}\phi(x)$ we find that (2.100) transforms to

$$-\beta\phi'' + \left(\frac{U'^2}{4\beta} - \frac{U''}{2}\right)\phi = \lambda\phi \tag{2.102}$$

where we have used (2.101). Equation (2.102) can be at once recognized to be of supersymmetric nature since we can write it as

$$\beta A^+ A\phi = \lambda\phi \tag{2.103}$$

where

$$A = \frac{\partial}{\partial x} + W(x)$$
$$A^+ = -\frac{\partial}{\partial x} + W(x) \tag{2.104}$$

and the superpotential $W(x)$ is related to $U(x)$ as $W(x) = \left(\frac{\partial U}{\partial x}\right)/2\beta$. The zero-eigenvalue of (2.a03) coresponds to $A\phi_0 = 0$ yielding $\phi_0 = Be^{-\frac{1}{2\beta}\int^x W(y)dy}$, B a constant. The supersymmetric partner to H_+ (the quantity β can be scaled appropriately) is given by $H_- \equiv \beta AA^+$.

The eigenvalue that controls the rate at which equilibrium is approached is the first excited eigenvalue E_1 of H_+ component. (Note the energy eigenvlaues of H_+ in increasing order are $0, E_1, E_2, \ldots$ while those of H_- are $E_1, E_2, \ldots$). E_1 is expected to be exponentially small since from qualitative considerations the potential depicts three minima and the probability of tunneling-transitions between different minima narrows the gap between the lowest and the first excited states exponentially. To evaluate E_1 it is to be noted that E_1 is the ground-state energy of H_-. Using suitable trial wave functions, E_1 can be determined variationally. Such a calculation also gives E_1 to be exponentially small as $\beta \to \infty$. For the derivation of Fokker-Planck equation and explanation of the variational estimate of E_1 see [88].

2.6 Superspace Formalism

An elegant description of SUSY can be made by going over [7,89,90] to the superspace formalism involving Grassmannian variables and then constructing theories based on superfields of such anticommuting variables. The simplest superspace contains a single Grassmann variable θ and constitutes what is known as $N = 1$ supersymmetric mechanics. The rule for the differentiation and integration of the Grassmann numbers is given as follows [91]

$$\frac{d}{d\theta}1 = 0,$$

$$\frac{d}{d\theta_i}\theta_j = \delta_{ij},$$

$$\frac{d}{d\theta_i}(\theta_k\theta_l) = \delta_{ij}\theta_l - \delta_{il}\theta_k \tag{2.105}$$

$$\int d\theta_i\theta_j = \delta_{ij},$$

$$\int d\theta_i = 0 \tag{2.106}$$

The above relations are sufficient to set up the underlying supersymmetric Lagrangian.

In the superspace spanned by the ordinary time variable t and anticommuting θ, we seea invariance of a differential line element under supersymmetric transformations parametrized by the Grassmann variable ϵ. It is easy to realize that under the combined transformations

$$\theta \to \theta' = \theta + \epsilon$$
$$t \to t' = t + i\epsilon\theta \tag{2.107}$$

the quantity $dt - i\theta d\theta$ goes into itself.

$$dt' - i\theta'd\theta' \to dt - i\theta d\theta \tag{2.108}$$

Note that the factor i is inserted in (2.108) to keep the line element real.

We next define a real, scalar superfield $\Phi(t,\theta)$ having a general form

$$\Phi(t,\theta) = q(t) + i\theta\psi(t) \tag{2.109}$$

where $\psi(t)$ is fermionic. Since we are dealing with a single Grassmann variable θ, the above is the most general representation of $\Phi(t, \theta)$. Writing

$$\delta\Phi = \Phi(t', \theta') - \Phi(t, \theta) \tag{2.110}$$

we can determine $\delta\Phi$ to be

$$\delta\Phi = i\epsilon\psi - i\epsilon\dot{q} \tag{2.111}$$

Comparison with (2.109) reveals the following transformations for q and ψ

$$\delta q = i\epsilon\psi, \quad \delta\psi = \epsilon\dot{q} \tag{2.112}$$

These point to a mixing of fermionic (bosonic) variables into the bosonic (fermionic) counterparts. Further $\delta\Phi$ can also be expressed as

$$\delta\Phi = \epsilon Q\Phi \tag{2.113}$$

where,

$$Q = \frac{\partial}{\partial\theta} + i\theta\frac{\partial}{\partial t} \tag{2.114}$$

Notice that $QQ\Phi = i\dot{q} - \theta\dot{\psi} = i\frac{\partial}{\partial t}\Phi$ implying that Q^2 on Φ gives the time-derivative. Since the generator of time-translation is the Hamiltonian we can express this result as

$$\{Q, Q\} = H \tag{2.115}$$

Identifying Q as the supersymmetric generator we see that (2.115) is in a fully supersymmetric form. Also replacing i by $-i$ in Q we can define another operator

$$D = \frac{\partial}{\partial\theta} - i\theta\frac{\partial}{\partial t} \tag{2.116}$$

which apart from being invariant under (2.107) gives $\{Q, D\} = 0$.

The Hamiltonian in (2.115) corresponds to that of a supersymmetric oscillator. To find the corresponding Lagrangian we notice that $D\Phi = i\psi - i\theta\dot{q}$, and as a result $D\Phi\dot{\Phi} = i\psi\dot{q} + \theta(\psi\dot{\psi} - i\dot{q}^2)$. We can therefore propose the following Lagrangian for $N = 1$ SUSY mechanics

$$L = \frac{i}{2}\int d\theta D\Phi\dot{\Phi} \tag{2.117}$$

Using (2.106), L can be reduced to $L = \frac{1}{2}\dot{q}^2 + \frac{i}{2}\psi\dot{\psi}$ which describes a free particle. It may be checked that δL yields a total derivative. In fact using (2.109) we find $\delta L = \frac{i\epsilon}{2}\frac{d}{dt}(\psi\dot{q})$.

In place of the scalar superfield $\Phi(t,\theta)$ if we had considered a 3-vector $\Phi(t,\theta)$ we would have been led to the nonrelativistic Pauli Hamiltonian for a quantized spin $\frac{1}{2}$ particle. If however, a 4-vector superfield $\Phi_\mu(t,\theta)$ was employed we would have gotten a relativistic supersymmetric version of the Dirac spin $\frac{1}{2}$ particle.

We now move on to a formulation of $N = 2$ supersymmetric mechanics which involves 2 Grassmannian variables. Let us call them θ_1 and θ_2. The relevant transformations for $N = 2$ case are

$$
\begin{aligned}
\theta_1 \to \theta_1' &= \theta_1 + \epsilon_1, \\
\theta_2 \to \theta_2' &= \theta_2 + \epsilon_2, \\
t \to t' &= t + i\epsilon_1\theta_1 + i\epsilon_2\theta_2
\end{aligned}
\tag{2.118}
$$

with ϵ_1 and ϵ_2 denoting the parameters of transformations. Obviously, under (2.118) the differential element $dt - i\theta_1 d\theta_1 - i\theta_2 d\theta_2$ remains invariant. For convenience let us adopt a set of complex representations $\theta, \bar{\theta} = (\theta_1 \mp i\theta_2)/\sqrt{2}$ and $\epsilon, \bar{\epsilon} = (\epsilon_1 \mp i\epsilon_2)/\sqrt{2}$. Note that $\theta^2 = 0 = \bar{\theta}^2$ and $\{\theta, \bar{\theta}\} = 0$. Similarly for ϵ and $\bar{\epsilon}$. Further, θ and $\bar{\theta}$ can be considered as complex conjugate to each other.

In the presence of 2 anticommuting variables θ and $\bar{\theta}$, the $N = 2$ superfield $\Phi(t, \theta, \bar{\theta})$ can be written as

$$
\Phi(t, \theta, \bar{\theta}) = q(t) + i\bar{\theta}\psi(t) + i\theta\bar{\psi}(t) + \theta\bar{\theta}A(t)
\tag{2.119}
$$

where $q(t)$ and $A(t)$ are real variables being bosonic in nature and $(\psi, \bar{\psi})$ are fermionic.

We can determine $\delta\Phi$ to be

$$
\delta\Phi = \left(\epsilon Q + \epsilon\bar{Q}\right)\Phi
\tag{2.120}
$$

where Q and $\bar{Q}$ are the generators of supersymmetric transformations

$$
\begin{aligned}
Q &= \frac{\partial}{\partial\bar{\theta}} + i\theta\frac{\partial}{\partial t} \\
\bar{Q} &= \frac{\partial}{\partial\theta} + i\bar{\theta}\frac{\partial}{\partial t}
\end{aligned}
\tag{2.121}
$$

Note that the supersymmetric transformations (2.118) induce the following transformations among $q(t), \psi(t), \overline{\psi}(t)$ and $A(t)$

$$
\begin{aligned}
\delta q &= i\epsilon\overline{\psi} + i\overline{\epsilon}\psi \\
\delta A &= \dot{\epsilon\psi} - \dot{\overline{\epsilon}\psi} \\
\delta\overline{\psi} &= -\overline{\epsilon}(\dot{q} + iA) \\
\delta\psi &= -\epsilon(\dot{q} - iA)
\end{aligned} \tag{2.122}
$$

Further the derivatives

$$
\begin{aligned}
D &= \frac{\partial}{\partial\overline{\theta}} - i\theta\frac{\partial}{\partial t} \\
\overline{D} &= \frac{\partial}{\partial\theta} - i\overline{\theta}\frac{\partial}{\partial t}
\end{aligned} \tag{2.123}
$$

anticommute with Q and $\overline{Q}$. These derivatives act upon Φ to produce

$$
\begin{aligned}
D\Phi &= i\psi - \theta A - i\theta\dot{q} + \theta\overline{\theta}\dot{\psi} \\
\overline{D}\Phi &= i\overline{\psi} + \overline{\theta}A - i\overline{\theta}\dot{q} - \theta\overline{\theta}\,\dot{\overline{\psi}}
\end{aligned} \tag{2.124}
$$

An educated guess for the $N = 2$ supersymmetric Lagrangian is

$$
L = \int d\theta d\overline{\theta}\left[\frac{1}{2}\overline{D}\Phi D\Phi - U(\Phi)\right] \tag{2.125}
$$

where $U(\Phi)$ is some function of Φ. Expanding $U(\Phi)$ as $U(\Phi) = U(0)$ $+\Phi U'(0) + \frac{|\Phi|^2}{2}U''(0) + \ldots$ where the derivatives are taken for $\theta = 0 = \overline{\theta}$, we find on using (2.124), $U(\Phi) = \theta\overline{\theta}(AU' + \overline{\psi}\psi U'') + \ldots$, where a prime denotes a derivative with respect to q. Carrying out the θ and $\overline{\theta}$ integrations we are thus led to

$$
L = \frac{1}{2}\dot{q}^2 + \frac{1}{2}A^2 - AU' + i\overline{\psi}\dot{\psi} - \overline{\psi}\psi U'' \tag{2.126}
$$

An immediate consequence of this L is that $\frac{\partial L}{\partial A} = 0$ (since L is independent of any $\dot{A}$ term) yielding $A = U'$. L can now be rearranged to be expressed as

$$
L = \frac{1}{2}\dot{q}^2 - \frac{1}{2}U'^2 + i\overline{\pi}\dot{\psi} - \overline{\psi}\psi U'' \tag{2.127}
$$

It is not difficult to see that if one uses the equations of motion [66]

$$\dot{\psi} = iU''\psi,$$
$$\dot{\overline{\psi}} = iU''\overline{\psi} \qquad (2.128)$$

L is found to possesi a conserved charge $\overline{\psi}\psi$ which is the fermion number operator N_F. Writing in (2.126) $\overline{\psi}\psi$ as $\frac{1}{2}(\overline{\psi}\psi - \psi\overline{\psi})$ and identifying the fermionic operators a and a^+ as the quantized versions of ψ and $\overline{\psi} = (\equiv \psi^+)$, respectively, [in (2.127) the variables ψ and $\overline{\psi}$ play the role of classical fermionic variables], one can make a transition to the Hamiltonian H of the system (2.127). H turns out to be

$$H = \frac{1}{2}p^2 + \frac{1}{2}U'^2 + \frac{1}{2}U''\sigma_3 \qquad (2.129)$$

where the momentum conjugate to ψ is clearly $-i\overline{\psi}$ while $\{\psi,\psi\} = 0 = \{\overline{\psi},\overline{\psi}\}$, $\{\psi,\overline{\psi}\} = 1$ and $(\psi,\overline{\psi})$ represented by $\frac{1}{2}\sigma_\mp$.

This H is of the same form as (2.26) if we make the identification $U' = -W$, the superpotential. Note that N_F becomes $\frac{1}{2}(1+\sigma_3)$. The above forms of the Lagrangian and Hamiltonian are the ones relevant to $N = 2$ supersymmetric quantum mechanics. This completes our discussion on the construction of the Lagrangians for $N = 1$ and 2 supersymmetric theories. In this connection it is interesting to note that the notion of seeking supersymmetric extensions has been successful in establishing superformulation of various algebras [92]. It has been possible [93] to relate the superconformal algebra to the supersymmetric extension of integrable systems such as the KdV equation [94-96]. Moreover a representative for the $SO(N)$ or $U(N)$ superconforma algebra has been found possible in terms of a free boson, N free fermions, and an accompanying current algebra [97].

It is also worth emphasizing that the properties characteristic of fermionic variables are crucial to the development of the supersymmetrization procedure. Noting that the fermionic quantities f_i, f_j form the basis of a Clifford algebra CL_{2n} given by

$$\{f_{+j}, f_{-l}\} = \delta_{il}I, \qquad (2.130)$$
$$\{f_{\pm j}, f_{\pm l}\} = 0 \qquad (2.131)$$

if we consider the replacement [98] of the rhs of (2.131) as $\delta_{jl}I \to \delta_{jl}I - 2O_{jl}$ with $O_{jl} = -O_{lj}, O_{jl}^+ = O_{jl}$, we are led to a Hamiltonian of the type (2.35) but inclusive of a spin-orbit coupling team $\sim$ $(x_j p_l - x_l p_j) O_{jl}$. The conformal invariances associated with the supersymmetrized harmonic oscillator have been judiciously exploited in [99-101] and the largest kinematical and dynamical invariance properties characterizing a higher dimensional harmonic oscillator system, in the framework of spin-orbit supersymmetrization, have also been studied. Related works [102] also include the "exotic" supersymmetric schemes'in two-space dimensions arising for each pair of integers v_+ and v_- yielding an $N = 2(v_+ + v_-)$ superalgebra in nonrelativistic Chem-Simons theory.

2.7   Other Schemes of SUSY

From (2.34) it can be easily verified that the commutator of the operators A and A^+ is proportional to the derivative of the superpotential

$$[A, A^+] = 2\frac{dW}{dx} \tag{2.132}$$

In this section we ask the question, whether we can impose some group structure upon A and A^+ in the framework of SUSY.

Consider the following representations of A and A^+ [103]

$$A = e^{iy}\left[k(x)\frac{\partial}{\partial x} - ik'(x)\frac{\partial}{\partial y} + U(x)\right]$$

$$A^+ = e^{-iy}\left[-k(x)\frac{\partial}{\partial x} - ik'(x)\frac{\partial}{\partial y} + U(x)\right] \tag{2.133}$$

where a prime denotes partial derivative with respect to x and $k(x)$, $U(x)$ are arbitrary functions of x. It is readily found that if we introduce an additional operator

$$A_3 = -i\frac{\partial}{\partial y} \tag{2.134}$$

A, A^+ and A_3 satisfy the algebra [104]

$$[A, A^+] = -2aA_3 - bI$$
$$[A_3, A] = A$$

$$[A_3, A^+] = -A^+ \tag{2.135a, b, c}$$

where I is the identity operator and a, b are appropriate functions of x

$$
\begin{aligned}
a &= [k'(x)]^2 - k(x)k''(x) \\
b &= 2[k'(x)U(x) - k(x)U'(x)]
\end{aligned}
\tag{2.136}
$$

The simultaneous presence of the functions $a(x)$ and $b(x)$ in (2.135a) is of interest. Without $a(x)$, (2.135a) reduces to (2.132). This is because $a = 0$ is consistent with $k = 1, U = W$, and $b = -W'$. On the other hand, the case $b = 0$ is associated with $U(x) = 0$. Clearly, the latter is a new direction which does not follow from Witten's model. Note that when $b = 0$, $k(x) = \sin x$, and $a = 1$ so A, A^+, and A_3 may be identified with the generators of the $SU(1,1)$ group [105-107].

Given the representations in (2.133), we can work out the modified components H'_+ and H'_- as follows

$$
\begin{aligned}
H'_+ &= \frac{1}{2}A^+ A \\
&= \frac{1}{2}\left[-k^2\frac{\partial^2}{\partial x^2} + ikk''\frac{\partial}{\partial y} - kU' + k'U \right. \\
&\quad \left. + \left(U - ik'\frac{\partial}{\partial y} \right)^2 - ik'^2\frac{\partial}{\partial y} \right]
\end{aligned}
\tag{2.137}
$$

$$
\begin{aligned}
H'_- &= \frac{1}{2}AA^+ \\
&= \frac{1}{2}\left[-k^2\frac{\partial^2}{\partial x^2} - ikk''\frac{\partial}{\partial y} + kU' - k'U \right. \\
&\quad \left. + \left(U - ik'\frac{\partial}{\partial y} \right)^2 + k'^2\frac{\partial}{\partial y} \right]
\end{aligned}
\tag{2.138}
$$

It should be stressed that the variable y is not to be confused with an extra spatial dimension and merely serves as an auxiliary parameter. This means that for a physical eigenvalue problem, the square of the modules of the eigenfunction must be independent of y.

The above model has been studied in [103] and also by Janussis et al. [108], Chuan [109], Beckers and Ndimubandi [110] and Samanta [111]. In [108], a two-term energy recurrence relation has been derived Wittin the Lie admissible formulation of Santilli's theory [112-113]. In [109] a set of coupled equations has been proposed, a particular class of which is in agreement with the results of [103]. In [110] connections of (2.133) have been sought with quantum deormation. Further, in [111], (2.137) and (2.138) have been successfully applied to a variety of physical systems which include the particle in a box problem, Morse potential, Coulomb potential, and the isotropic oscillator potential.

Finally, let us remark that other extensions of the algebraic approach towards SUSYQM have also appeared (see Pashnev [114]) for $N = 2, 3$. Moreover, Verbaarschot et al. have calculated the large order behaviour of $N = 4$ SUSYQM using a perturbative expansion [115]. The pattern of behavior has been found to be of the same form as models with two supersymmetrics. Akulov and Kudinov [116] have considered the possibility of enlarging SUSYQM to any N by expressing the Hamiltonian as the sum of irreducible representations of the symmetry group S_N. To set up a working scheme certain compatibility conditions arise by requiring the representations to be totally symmetric and to satisfy a superalgebra. Very recently, Znojil et al. [117] have constructed a scheme of SUSY using nonhermitean operators. Its representation spere is spanned by bound states with PT symmetry but yields real energies.

2.8   References

[1] P.A.M. Dirac, *The Principles of Quantum Mechanics*, 4th ed., Clarendon Press, Oxford, 1958.

[2] L.I. Schiff, *Quantum Mechanics*, 3rd ed., McGraw-Hill, New York, 1968.

[3] S. Gasiorowicz, *Quantum Physics*, John Wiley & Sons, New York, 1974.

[4] O.L. de Lange and R.E. Raab, *Operator Methods in Quantum Mechanics*, Clarendon Press, Oxford, 1991.

[5] R.J. Glauber, in *Recent Developments in Quantum Optics*, R. Inguva, Ed., Plenum Press, New York, 1993.

[6] E. Fermi, *Notes on Quantum Mechanics*, The University of Chicago Press, Chicago, 1961.

[7] F. Ravndahl, Proc CERN School of Physics, 302, 1984.

[8] E. Witten, *Nucl. Phys.*, **B188**, 513, 1981.

[9] E. Witten, *Nucl. Phys.*, **B202**, 253, 1982.

[10] A.I. Vainshtein, A.V. Smilga, and M.A. Shifman, *Sov. Phys. JETP.*, **67**, 25, 1988.

[11] A. Arai, *Lett. Math. Phys.*, **19**, 217, 1990.

[12] P. Salomonson and J.W. van Holten, *Nucl. Phys.*, **B196**, 509, 1982.

[13] M. Clandson and M. Halpern, *Nucl. Phys.*, **B250**, 689, 1985.

[14] F. Cooper and B. Freedman, *Ann. Phys.*, **146**, 262, 1983.

[15] G. Parisi and N. Sourlas, *Nucl. Phys.*, **B206**, 321, 1982.

[16] S. Cecotti and L. Girardello, *Ann. Phys.*, **145**, 81, 1983.

[17] L.F. Urrutia and E. Hernandez, *Phys. Rev. Lett.*, **51**, 755, 1983.

[18] M. Bernstein and L.S. Brown, *Phys. Rev. Lett.*, **52**, 1983, 1984.

[19] H.C. Rosu, *Phys. Rev.*, **E56**, 2269, 1997.

[20] E. Gozzi, *Phys. Rev.*, **D30**, 1218, 1984.

[21] E. Gozzi, *Phys. Rev.*, **D33**, 584, 1986.

[22] J. Húrby, *Czech J. Phys.*, **B37**, 158, 1987.

[23] J. Maharana and A. Khare, *Nucl. Phys.*, **B244**, 409, 1984.

[24] R. Akhoury and A. Comtet, *Nucl. Phys.*, **B245**, 253, 1984.

[25] D. Boyanovsky and R. Blankenbekler, *Phys. Rev.*, **D30**, 1821, 1984.

[26] L.E. Gendenshtein, *JETP Lett.*, **38**, 356, 1983.

[27] A.A. Andrianov, N.V. Borisov, and M.V. Ioffe, *Phys. Lett.*, **A105**, 19, 1984.

[28] M.M. Nieto, *Phys. Lett.*, **B145**, 208, 1984.

[29] C.V. Sukumar, *J. Phys. A. Math. Gen.*, **18**, L57, 2917, 2937, 1985.

[30] C.V. Sukumar, *J. Phys. A. Math. Gen.*, **A20**, 2461, 1987.

[31] V.A. Kostelecky, M.M. Nieto, and D.R. Truax, *Phys. Rev.*, **D32**, 2627, 1985.

[32] V.A. Kosteleeky and M.M. Nieto, *Phys. Rev. Lett.*, **53**, 2285, 1984.

[33] V.A. Kosteleeky and M.M. Nieto, *Phys. Rev. Lett.*, **56**, 96, 1986.

[34] V.A. Kosteleeky and M.M. Nieto, *Phys. Rev.* **A32**, 1293, 1985.

[35] E. D'Hoker and L. Vinet, *Phys. Lett.*, **B137**, 72, 1984.

[36] C.A. Blockley and G.E. Stedman, *Eur. J. Phy.*, **6**, 218, 1985.

[37] P.D. Jarvis and G. Stedman, *J. Phys. A. Math. Gen.*, **17**, 757, 1984.

[38] M. Baake, R. Delbourgo, and P.D. Jarvis, *Aust. J. Phys.*, **44**, 353, 1991.

[39] de Crombrugghe and V. Rittenberg, *Ann. Phys.*, **151**, 99, 1983.

[40] D. Lancaster, *Nuovo Cim*, **A79**, 28, 1984.

[41] S. Fubini and E. Rabinovici, *Nucl. Phys.*, **B245**, 17, 1984.

[42] H. Yamagishi, *Phys. Rev.*, **D29**, 2975, 1984.

[43] H. Ui and G. Takeda, *Prog. Theor. Phys.*, **72**, 266, 1984.

[44] H. Ui, *Prog. Theor. Phys.*, **72**, 813, 1984.

[45] D. Sen, *Phys. Rev.*, **D46**, 1846, 1992.

[46] L.E. Gendenshtein and I.V. Krive, *Sov. Phyi. Usp.*, **28**, 645, 1985.

[47] R.W. Haymaker and A.R.P. Rau, *Am. J. Phys.*, **54**, 928, 1986.

[48] W. Kwong and J.L. Rosner, *Prog. Theor. Phys.* (Supp.), **86**, 366, 1986.

[49] A. Lahiri, P.K. Roy, and B. Bagchi, *Int. J. Mod. Phys.*, **A5**, 1383, 1990.

[50] B. Roy, P. Roy, and R. Roychoudhury, *Fortsch Phys.*, **39**, 211, 1991.

[51] G. Levai, Lecture Notes in Physics, **427**, 127, Springer, Berlin, 1993.

[52] F. Cooper, A. Khare, and U. Sukhatme, *Phys. Rep.*, **251**, 267, 1995.

[53] G. Junker, *Supersymmetric Methods in Quantum and Statistical Physics,* Springer, Berlin, 1996 and references therein.

[54] A. Comtet, A.D. Bandrauk, and D.K. Campbell, *Phys. Lett.*, **B150**, 159, 1985.

[55] A. Khare, *Phys. Lett.*, **B161**, 131, 1985.

[56] J. Mañes and B. Zumino, *Nucl. Phys.*, **B270**, 651, 1986.

[57] K. Raghunathan, M. Seetharaman, and S.S. Vasan, *Phys. Lett.*, **188**, 351, 1987.

[58] S.S. Vasan, M. Seetharaman, and K. Raghunathan, *J. Phys. A. Math. Gen.*, **21**, 1897, 1988.

[59] R. Dutt, A. Khare, and U. Sukhatme, *Phys. Lett.*, **B181**, 295, 1986.

[60] R. Dutt, A. Gangopadhyaya, A. Khare, A. Pagnamenta, and U. Sukhatme, *Phys. Rev.*, **A48**, 1845, 1993.

[61] D. DeLaney and M.M. Nieto, *Phys. Lett.*, **B247**, 301, 1990.

[62] B. Chakrabarti and T.K. Das, *Phys. Rev.*, **A60**, 104, 1999.

[63] A. Inomata and G. Junker, *Phys. Rev.*, **A50**, 3638, 1994.

[64] B. Bagchi, S.N. Biswas, and R. Dutt, Parametrizing the partner Hamiltonians in SUSYQM, preprint, 1995.

[65] P.G. Leach, *Physica*, **D17**, 331, 1985.

[66] M.A. Shiftman, ITEP Lectures on Particle Physics and Field Theory, *World Scientific*, **62**, 301, 1999.

[67] H. Ui, *Prog. Theor. Phys.*, **72**, 192, 1984.

[68] R. Jackiw, *Phys. Rev.*, **D29**, 2375, 1984.

[69] R.J. Hughes, V.A. Kosteleky, and M.M. Nieto, *Phys. Lett.*, **B171**, 226, 1986.

[70] R.J. Hughes, V.A. Kostelecky, and M.M. Nieto, *Phys. Rev.*, **D34**, 1106, 1986.

[71] J. Beckers and N. Debergh, *Phys. Rev.*, **D42**, 1255, 1990.

[72] O. Castaños, A. Frank, R. López, and L.F. Urrutia, *Phys. Rev.*, **D43**, 544, 1991.

[73] C.V. Sukumar, *J. Phys. A. Math. Gen.*, **18**, L697, 1985.

[74] C.J. Lee, *Phys. Rev.*, **A50**, 2053, 1994.

[75] L. O'Raifeartaigh, Lecture notes on supersymmetry, *Comm. Dublin Inst. for Adv. Studies*, Series A, **22**, 1975.

[76] I. Kay and H.E. Moses, *J. Appl. Phys.*, **27**, 1503, 1956.

[77] I.M. Gel'fand and B.M. Levitan, *Am Math. Soc. Trans.*, **1**, 253, 1955.

[78] V.A. Marchenko, Dokl Akad Nauk SSSR, **104**, 695, 1955.

[79] W. Kwong, H. Riggs, J.L. Rosner, and H.B. Thacker, *Phys. Rev.*, **D39**, 1242, 1989.

[80] C. Quigg, H.B. Thacker, and J.L. Rosner, *Phys. Rev.*, **D21**, 234, 1980.

[81] C. Quigg and J.L. Rosner, *Phys. Rev.*, **D23**, 2625, 1981.

[82] P. Moxhay, J.L. Rosner, and C. Quigg, *Phys. Rev.*, **D23**, 2638, 1981.

[83] W. Kwong and J.L. Rosner, *Phys. Rev.*, **D38**, 279, 1988.

[84] A. Anderson, *Phys. Rev.*, **A43**, 4602, 1991.

[85] J. Schonfeld, W. Kwong, J.L. Rosner, C. Quigg, and H.B. Thacker, *Ann. Phys.*, **128**, 1, 1990.

[86] B. Bagchi, *Int. J. Mod. Phys.*, **A5**, 1763, 1990.

[87] N. Wax Ed, *Selected Papers on Noise and Stochastic Processes*, Dover, NY, 1954.

[88] J.K. Bhattacharjee, *Statistical Physics: Equilibrium and Non-Equilibrium Aspect*, Allied, New Delhi, 1997.

[89] J. Wess, and B. Zumino, *Supersymmetry and Supergravity*, Princeton University Press, Princeton, NJ, 1983.

[90] P.G.O. Freund, *Introduction to Supersymmetry, Cambridge Monographs on Mathematical Physics*, Cambridge University Press, Cambridge, 1986.

[91] F.A. Berezin, *The Method of Second Quantization*, Academic Press, New York, 1966.

[92] M. Chaichian and P. Kullish, *Phys. Lett.*, **B183**, 169, 1987.

[93] P. Mathieu, *Phys. Lett.*, **B203**, 287, 1988 and references therein.

[94] J.L. Gervais and A. Neveu, *Nucl. Phys.*, **B209**, 125, 1982.

[95] J.L. Gervais, *Phys. Lett.*, **B160**, 277, 279, 1985.

[96] P. Mathieu, *Jour. Math. Phys.*, **29**, 2499, 1988.

[97] P. Mathieu, *Phys. Lett.*, **B218**, 185, 1989.

[98] N. Debergh, *J. Phys. A. Math. Gen.*, **24**, 147, 1991.

[99] J. Beckers, D. Dehin, and V. Hussin, *J. Phys. A. Math. Gen.*, **21**, 651, 1988.

[100] J. Beckers, D. Dehin, and V. Hussin, *J. Phys. A. Math. Gen.*, **20**, 1137, 1987.

[101] J. Beckers, N. Debergh, V. Hussin, and A. Sciarrino, *J. Phys. A. Math. Gen.*, **23**, 3647, 1990.

[102] C. Duval and P.A. Horváthy, *J. Math. Phys.*, **35**, 2516, 1994 and references therein.

[103] A. Lahiri, P. Roy, and B. Bagchi, *Nuovo Cim*, **100A**, 797, 1988.

[104] S. Kais and R.D. Levine, *Phys. Rev.*, **A34**, 4615, 1986.

[105] A.B. Balantekin, *Ann. Phys.*, **164**, 277, 1985.

[106] J. Wu and Y. Alhassid, *J. Math. Phys.*, **31**, 557, 1990.

[107] M.J. Englefield and C. Quesne, *J. Phys. A. Math. Gen.*, **24**, 3557, 1991.

[108] A. Janussis, I. Tsohantzis, and D. Vavougios, *Nuovo Cim*, **105B**, 1171, 1990.

[109] C.X. Chuan, *J. Phys. A. Math. Gen.*, **23**, L659, 1990.

[110] J. Beckers and J. Ndimubandi, *Phys. Scripta*, **54**, 9, 1996.

[111] K. Samanta, *Int. J. Theor. Phys.*, **32**, 891, 1993.

[112] R. Santilli, *Hadronic. J.*, **2**, 1460, 1979.

[113] P. Srivastava, *Lett. Nuovo Cim.*, **15**, 588, 1976.

[114] A. Pashnev, *Theor. Math. Phys.*, **96**, 311, 1986.

[115] J.J.M. Verbaarschot, P. West, and T.T. Wu, *Phys. Lett.*, **B240**, 401, 1990.

[116] V. Akulov and M. Kudinov, *Phys. Lett.*, **B460**, 365, 1999.

[117] M. Znojil, F. Cannata, B. Bagchi, and R. Roychoudhury, *Phys. Lett.*, **B483**, 284, 2000.

CHAPTER 3

Supersymmetric Classical Mechanics

3.1 Classical Poisson Bracket, its Generalizations

In classical mechanics we encounter the notion of Poisson brackets in connection with transformations of the generalized coordinates and generalizaed momenta that leave the form of Hamilton's equations of motion unchanged [1-3]. Such transformations are called canonical and the main property of the Poisson bracket is its invariance with respect to the canonical transformations. In terms of the generalized coordinates $q_1, q_2, \ldots q_n$ and generalized momenta $p_1, p_2, \ldots, p_n$ the Poisson bracket in classical mechanics is defined by

$$\{f, g\} = \sum_{j=1}^{n} \left(\frac{\partial f}{\partial q_j} \frac{\partial g}{\partial p_j} - \frac{\partial f}{\partial p_j} \frac{\partial g}{\partial q_j} \right) \tag{3.1}$$

for any pair of functions $f \equiv f(q_1, q_2, \ldots q_n; p_1, p_2, \ldots p_n; t)$ and $g \equiv g(q_1, q_2, \ldots q_n; p_1, p_2, \ldots p_n; t)$.

Recall that whereas the Lagrangian in classical mechanics is known in terms of the generalized coordinates (q_i), the generalized velocities $(\dot{q}_i)$, and time (t), namely $L = L(q_1, q_2, \ldots q_n; \dot{q}_1, \dot{q}_2, \ldots \dot{q}_n, t)$, the corresponding Hamiltonian is given in terms of the generalized coordinates (q_i), generalized momenta (p_i), and time (t), namely

$H = H(q_1, q_2, \ldots q_n; p_1, p_2, \ldots p_n; t)$ where $p_i = \frac{\partial L}{\partial \dot{q}_i}, i = 1, 2, \ldots n$. The relationship between the Lagrangian and Hamiltonian is provided by the Legendre transformation $H = \sum_{i=1}^{n} p_i \dot{q}_i - L$ and Hamilton's canonical equations of motion are obtained by varying both sides of it

$$\frac{\partial H}{\partial p_i} = \dot{q}_i$$

$$\frac{\partial H}{\partial q_i} = -\dot{p}_i$$

$$\frac{\partial H}{\partial t} = -\frac{\partial L}{\partial t} \tag{3.2}$$

Relations (3.2) prescribe a set of $2n$ first-order differential equations for the $2n$ variables (q_i, p_i). In contrast Lagrange's equations involve n second-order differential equations for the n generalized coordinates

$$\frac{d}{dt}\left(\frac{\partial L}{\partial \dot{q}_i}\right) = \frac{\partial L}{\partial q_i} \quad i = 1, 2, \ldots n \tag{3.3}$$

For a given transformation $(q_i, p_i) \rightarrow (Q_i, P_i)$ to be canonical we need to have

$$\{Q_i, Q_j\} = 0, \{P_i, P_j\} = 0,$$

$$\{Q_i, P_j\} = \delta_{ij} \tag{3.4}$$

These conditions are both necessary and sufficient. Often (3.4) is used as a definition for the canonically conjugate coordinates and momenta. Some obvious properties of the Poisson brackets are

anti-symmetry: $\qquad \{f, g\} = -\{g, f\}, \{f, c\} = 0 \qquad$ (3.5a)

linearity: $\qquad \{f_1 + f_2, g\} = \{f_1, g\} + \{f_2, g\} \qquad$ (3.5b)

chain-rule: $\qquad \{f_1 f_2, g\} = f_1\{f_2, g\} + \{f_1, g\}f_2 \qquad$ (3.5c)

Jacobi identity: $\{f, \{g, h\}\} + \{g, \{h, f\}\} + \{h, \{f, g\}\} = 0$ (3.5d)

where c is a constant and the functions involved are known in terms of generalized coordinates, momenta, and time.

The transition from classical to quantum mechanics is formulated in terms of the commutators from the classical Poisson bracket relations. Indeed it can be readily checked that the commutator of two operators satisfies all the properties of the Poisson bracket summarized in (3.5). The underlying fundamental commutation relation in quantum mechanics being $[x, p] = i\hbar$, the classical Poisson bracket may be viewed as an outcome of the following limit on the commutator

$$\lim_{\hbar \to 0} \frac{\left[\widehat{f}, \widehat{g}\right]}{i\hbar} = \{f, g\} \tag{3.6}$$

where $\left[\widehat{f}, \widehat{g}\right]$ stands for the commutator of the two operators $\widehat{f}$ and $\widehat{g}$.

It is also possible to work on the $\hbar \to 0$ limit (that is, the classical limit) of the quantum theory involving fermionic degrees of freedom [4]. This requires the corresponding classical Lagrangian to have in addition to the usual generalized coordinates and velocities, anti-commuting variables and their time-derivatives. We must therefore distinguish, at the quantum level, between those operators which are even or odd under a permutation operator P

$$P^{-1} \widehat{A} P = (-1)^{\pi(\widehat{A})} \widehat{A} \tag{3.7}$$

where $\widehat{A}$ is some operator and $\pi(\widehat{A})$ is defined by

$$\begin{aligned} \pi(\widehat{A}) &= 0 \text{ if } \widehat{A} \text{ is even} \\ &= 1 \text{ if } \widehat{A} \text{ is odd} \end{aligned} \tag{3.8}$$

An even operator transforms even (odd) states into even (odd) states while an odd operator transforms even (odd) states into odd (even) states. In keeping with the properties of an ordinary commutator, which as stated before are the same as those of the Poisson brackets outlined in (3.5), we can think of a generalized commutator (also called the generalized Dirac bracket) as being the one which is obtained by taking into account the evenness or oddness of an operator. Thus setting $\pi(\widehat{A}) = a$, $\pi(\widehat{B}) = b$, and $\pi(\widehat{C}) = c$, the generalized Dirac bracket $\left[\widehat{A}, \widehat{B}\right]$ is defined such that the following properties hold

anti-symmetry: $[\widehat{A}, \widehat{B}] = -(-1)^{ab}[\widehat{B}, \widehat{A}]$

chain-rule: $[\widehat{A}, \widehat{B}\widehat{C}] = [\widehat{A}, \widehat{B}]\widehat{C} + (-1)^{ab}\widehat{B}[\widehat{A}, \widehat{C}]$

linearity: $[\widehat{A}, \widehat{B} + \widehat{C}] = [\widehat{A}, \widehat{B}] + [\widehat{A}, \widehat{C}]$

Jacobi identity: $[\widehat{A}, [\widehat{B}, \widehat{C}]] + (-1)^{ab+ac}[\widehat{B}[\widehat{C}, \widehat{A}]]$

$$+ (-1)^{ca+cb}[\widehat{C}, [\widehat{A}, \widehat{B}]] = 0 \quad (3.9a, b, c, d)$$

Clearly, these properties are the analogs of the corresponding ones stated in (3.5). In the absence of any fermionic degrees of freedom it is evident that (3.9) reduces to the usual properties of the commutators.

The chain-rule allows us to recognize $[\widehat{A}, \widehat{B}]$ as

$$[\widehat{A}, \widehat{B}] = \widehat{A}\widehat{B} - (-1)^{ab}\widehat{B}\widehat{A} \tag{3.10}$$

which implies that $[\widehat{A}, \widehat{B}]$ plays the role of an anti-commutator when $\widehat{A}$ and $\widehat{B}$ are odd but a commutator otherwise

$$
\begin{aligned}
[\widehat{A}, \widehat{B}] &= \widehat{A}\widehat{B} + \widehat{B}\widehat{A} \quad \widehat{A} \text{ and } \widehat{B} \text{ odd} \\
&= \widehat{A}\widehat{B} - \widehat{B}\widehat{A} \quad \text{otherwise}
\end{aligned}
\tag{3.11}
$$

With the definition (3.10) and the use of the linearity and chain-rule properties, the Jacobi identity (3.9d) can be seen to hold.

To derive (3.10) it is instructive to evaluate $[\widehat{A}\widehat{B}, \widehat{C}\widehat{D}]$, where $\widehat{C}$ and $\widehat{D}$ are also operators. Applying (3.9b) in two different ways, we get

$$
\begin{aligned}
[\widehat{A}\widehat{B}, \widehat{C}\widehat{D}] &= [\widehat{A}\widehat{B}, \widehat{C}]\widehat{D} + (-1)^{\pi(\widehat{A}\widehat{B})\pi(\widehat{C})}\widehat{C}[\widehat{A}\widehat{B}, \widehat{D}] \\
&= [\widehat{A}\widehat{B}, \widehat{C}]\widehat{D} + (-1)^{(a+b)c}\widehat{C}[\widehat{A}\widehat{B}, \widehat{D}]
\end{aligned}
\tag{3.12}
$$

where we have used $\pi(\widehat{A}\widehat{B}) = \pi(\widehat{A}) + \pi(\widehat{B}) = a + b$ and applied tee chain-rule on $\widehat{C}\widehat{D}$. Next using (3.9a) and once again (3.9b) we arrive at

$$[\widehat{A}\widehat{B}, \widehat{C}\widehat{D}] = (-1)^{bc}[\widehat{A}, \widehat{C}]\widehat{B}\widehat{D} + \widehat{A}[\widehat{B}, \widehat{C}]\widehat{D} + (-1)^{ac+bc+bd}$$

$$\widehat{C}[\widehat{A}, \widehat{D}]\widehat{B} + (-1)^{ac+bc}\widehat{C}\widehat{A}[\widehat{B}, \widehat{D}]$$

$$\tag{3.13a}$$

Applying now (3.9b) on $\widehat{A}\widehat{B}$ we have

$$[\widehat{A}\widehat{B}, \widehat{C}\widehat{D}] = (-1)^{bc+bd}[\widehat{A}, \widehat{C}]\widehat{D}\widehat{B} + \widehat{A}[\widehat{B}, \widehat{C}]\widehat{D} + (-1)^{ac+bc+bd}$$

$$\widehat{C}[\widehat{A}, \widehat{D}]\widehat{B} + (-1)^{bc}\widehat{A}\widehat{C}[\widehat{B}, \widehat{D}]$$

$$\tag{3.13b}$$

Since (3.13a) and (3.13b) are equivalent representations of $[\widehat{A}\widehat{B}, \widehat{C}\widehat{D}]$ we get on equating them

$$[\widehat{A}, \widehat{C}] \left\{ \widehat{B}\widehat{D} - (-1)^{bd} \widehat{D}\widehat{B} \right\} = \left\{ \widehat{A}\widehat{C} - (-1)^{ac} \widehat{C}\widehat{A} \right\} [\widehat{B}, \widehat{D}] \quad (3.14)$$

(3.14) implies that the generalized bracket $[X, Y]$ involving two operators X and Y can be identitified as

$$[X, Y] = XY - (-1)^{\pi(X)\pi(Y)} YX \quad (3.15)$$

It is obvious that (3.15) is consistent with (3.11).

The generalized bracket (3.15) gives way to a formulation of the quantized rule

$$\lim_{\hbar \to 0} \frac{[X, Y]}{i\hbar} = \{X, Y\} \quad (3.16)$$

where $[X, Y]$ has been defined according to (3.15) and $\{X, Y\}$ stands for the corresponding classical Poisson bracket. Note that the classical system possesses not only commuting variables such as the q's and p's but also additional anti-commuting degrees of freedom. So the Poisson bracket in (3.16) is to be looked upon in a generalized sense [5-12].

3.2 Some Algebraic Properties of the Generalized Poisson Bracket

Pseudomechanics or pseudoclassical mechanics as named by Casalbuoni [5] is concerned with classical systems consisting of anti-commuting as well as c-number variables in the form of coordinates and momenta. Let θ_α's be a set of anti-commuting or Grassmann variables in addition to the coordinates q_i's. Then the pseudoclassical Lagrangian can be written as

$$L \equiv L\left(q_i, \dot{q}_i, \theta_\alpha, \dot{\theta}_\alpha\right) \quad (3.17)$$

We assumt for simplicity that L is not explicity dependent upon the time variable t. The corresponding Hamiltonian would be a function of even (bosonic) variables (q_i, p_i) and odd (fermionic) variables $(\theta_\alpha, \pi_\alpha)$ where p_i and π_α are the corresponding canonical momenta to the coordinates:

$$H \equiv H(q_i, \theta_\alpha, p_i, \pi_\alpha) \quad (3.18)$$

To develop a canonical formalism we need to impose upon the coordinates and momenta the conditions (3.4), namely

$$\begin{aligned}
\{Q_i, Q_j\} &= 0, \ \{P_i, P_j\} = 0 \\
\{Q_i, P_j\} &= \delta_{ij}
\end{aligned} \tag{3.19}$$

but here Q and P denote collectively the coordinates (q_i, θ_α) and the momenta (p_i, π_α).

To deal with the odd variables it is necessary to identify properly the processes of left and right differentiation. At the pure classical level where we deal with even variables only (like coordinates and momenta), such a distinction is not relevant. However, in a pseudo-classical system in which the dynamical variable X is a function of Q and P, its differential needs to be specified as [12]

$$\delta X(Q, P) = X_{,Q} dQ + dP \partial_P X \tag{3.20}$$

where a right-derivative is taken with respect to the coordinates Q and a left-derivative with respect to the momenta P. By accounting for the permutations correctly we can write

$$\partial_Q X = (-1)^{\pi(Q)[\pi(Q)+\pi(X)]} X_{,Q} \tag{3.21}$$

A consequence of (3.21) is that

$$\begin{aligned}
\partial_\theta O &= O_{,\theta}, \ \partial_\pi O = O_{,\pi} \\
\partial_\theta E &= -E_{,\theta}, \ \partial_\pi E = -E_{,\pi} \\
\partial_q O &= O_{,q}, \ \partial_p O = O_{,p} \\
\partial_q E &= E_{,q}, \ \partial_p E = E_{,p}
\end{aligned} \tag{3.22}$$

where O and E denote odd and even variable respectively.

It is clear from (3.20) that the canonical momenta are to be defined as $P = L_{,\dot{Q}}$ implying that since the Lagrangian is an even function of the underlying variables we should have

$$p_i = \frac{\partial L}{\partial \dot{q}_i}, \ \pi_\alpha = -\frac{\partial L}{\partial \dot{\theta}_\alpha} \tag{3.23}$$

with $\{\pi^\alpha, \theta_\beta\} = 0$, $\alpha \neq \beta$.

To derive the generalized Hamilton's equation of motion we set up a Legendre transformateon from the classical analogy

$$H = \sum_i p_i \dot{q}_i + \sum_\alpha \pi_\alpha \dot{\theta}_\alpha - L \qquad (3.24)$$

Varying $H(q_i, p_i, \theta_\alpha, \pi_\alpha)$ and keeping in mind the rules (3.20) and (3.21), the equations of motion emerge as

$$
\begin{aligned}
\dot{q}_i &= \frac{\partial H}{\partial p_i}, \quad \dot{p}_i = -\frac{\partial H}{\partial q_i} \\
\dot{\theta}_\alpha &= \frac{\partial H}{\partial \pi_\alpha}, \quad \dot{\pi}_\alpha = \frac{\partial H}{\partial \theta_\alpha}
\end{aligned}
\qquad (3.25)
$$

Noting that the equation of motion of a dynamical variable X is given in terms of the Poisson bracket as $\frac{dX}{dt} = \frac{\partial X}{\partial t} + \{X, H\}$ and $\{X, H\}$ is defined according to

$$\{X, H\} = X_{,Q} \partial_P H - H_{,Q} \partial_P X \qquad (3.26)$$

[where we have followed (3.1) but made a distinction between the left and right derivatives], it is trivial to check using (3.22) that $\{\theta, H\} = \dot{\theta}$ and $\{\pi, H\} = \dot{\pi}$.

More generally, the generalized Poisson bracket for various cases of even and odd variables may be summarized as follows

$$\{E_1, E_2\} = \left(\frac{\partial E_1}{\partial q} \frac{\partial E}{\partial p} - \frac{\partial E_2}{\partial q} \frac{\partial E_1}{\partial p} \right) + \left(-\frac{\partial E_1}{\partial \theta} \frac{\partial E_2}{\partial \pi} + \frac{\partial E_2}{\partial \theta} \frac{\partial E_1}{\partial \pi} \right)$$

$$\{E, O\} = \left(\frac{\partial E}{\partial q} \frac{\partial O}{\partial p} - \frac{\partial O}{\partial q} \frac{\partial E}{\partial p} \right) - \left(\frac{\partial E}{\partial \theta} \frac{\partial O}{\partial \pi} + \frac{\partial O}{\partial \theta} \frac{\partial E}{\partial \pi} \right)$$

$$\{O, E\} = \left(\frac{\partial O}{\partial q} \frac{\partial E}{\partial p} - \frac{\partial E}{\partial q} \frac{\partial O}{\partial p} \right) + \left(\frac{\partial O}{\partial \theta} \frac{\partial E}{\partial \pi} + \frac{\partial E}{\partial \theta} \frac{\partial O}{\partial \pi} \right)$$

$$\{O_1, O_2\} = \left(\frac{\partial O_1}{\partial q} \frac{\partial O_2}{\partial p} + \frac{\partial O_2}{\partial q} \frac{\partial O_1}{\partial p} \right) + \left(\frac{\partial O_1}{\partial \theta} \frac{\partial O_2}{\partial \pi} + \frac{\partial O_2}{\partial \theta} \frac{\partial O_1}{\partial \pi} \right)$$

$$(3.27a, b, c, d)$$

An interesting feature with the structure of (3.27) is that the canonical relations (3.19) between the coordinates and momenta are automatically preserved. This enables us to derive a classical version of the supersymmetric Lagrangian in a straightforward manner.

Finally, the classical $\hbar \to 0$ limit of the quantization rule (3.16) may be written down with respect to the even and odd operators by making use of (3.8) and (3.10).

$$\begin{aligned}
[E_1, E_2]_- &= i\hbar\{E_1, E_2\} \\
[O, E]_- &= i\hbar\{O, E\} \\
[O_1, O_2]_+ &= i\hbar\{O_1, O_2\}
\end{aligned} \qquad (3.28)$$

where the right hand side denotes the generalized Poisson bracket with respect to both commuting and anti-commuting sets of variables and where $-$ and $+$ in the left hand side corresponds to the commutator and anti-commutator, respectively. It may be remarked that from (3.27b) and (3.27c) we have $\{O, E\} = -\{E, O\}$. The respective expressi,n for the Poisson bracket in (3.28) are those given by (3.27a), (3.27c) and (3.27d). We notice that only the odd-odd operators are quantized with respect to the anti-commutator while the remaining ones are quantized with respect to the commutator.

3.3   A Classical Supersymmetric Model

We now seek the classical supersymmetric Hamiltonian in the form

$$H_{Scl} = \{Q, Q^+\} \qquad (3.29)$$

with $\{Q, Q\} = \{Q^+, Q^+\} = 0$. Utilizing the Hamiltonian's equation of motion in Poisson bracket notation we have

$$\begin{aligned}
\dot{Q} &= \{Q, H_{Scl}\} \\
&= \{Q, \{Q, Q^+\}\} \\
&= -\{Q, \{Q, Q^+\}\} \\
&= -\dot{Q}
\end{aligned} \qquad (3.30)$$

where Jacobi identity has been used. So $\dot{Q} = 0$ and similarly $\dot{Q}^+ = 0$. These give at once

$$\{Q, H_{Scl}\} = \{Q^+, H_{Scl}\} = 0 \qquad (3.31)$$

implying that the conservation of Q and Q^+ is in-built in (3.29).

We can also write down explicit representations for Q and Q^+ by setting

$$Q = \frac{1}{\sqrt{2}}A\theta, \quad Q^+ = \frac{1}{\sqrt{2}}A^*\pi, \quad A = W + ip \qquad (3.32)$$

Note that since $\{\theta, \pi\} = 1$ the expressions (3.32) are just the classical analogs of the corresponding quantum quantities. From (3.29), (3.27d) and (3.27a) we have

$$
\begin{aligned}
H_{Scl} &= \frac{1}{2}\{A, A^*\}\theta\pi + \frac{1}{2}AA^*\{\theta, \pi\} \\
&= \frac{1}{2}p^2 + \frac{1}{2}W^2 - iW'\theta\pi \qquad (3.33)
\end{aligned}
$$

Here the potential $V_{Scl} = \frac{1}{2}W^2 - iW'\theta\pi$ matches with the one obtained from the classical limit of the SUSYQM Lagrangian given in the previous chapter. To see this we rewrite (2.127) up to a total derivative as

$$
\begin{aligned}
L_{Scl} &= \frac{1}{2}\dot{x}^2 - \frac{1}{2}U'^2 - \psi^+\psi U'' + \frac{i}{2}\left(\psi^+\dot{\psi} - \dot{\psi}^+\psi\right) \\
&= \frac{1}{2}\dot{x}^2 + \frac{i}{2}\left(\psi^+\dot{\psi} - \dot{\psi}^+\psi\right) - V \qquad (3.34)
\end{aligned}
$$

where

$$V = \frac{1}{2}U'^2 + \psi^+\psi U'' \qquad (3.35)$$

and an overhead dot stands for a time-derivative. Now from (2.127) the canonical momenta π for ψ is $i\psi^+$ which when substituted in V gives $V = \frac{1}{2}U'^2 + iU''\psi\pi$. We thus recover V_{Scl} if we identify $W = -U'$ and note that ψ plays the role θ.

To complete our discussion on the classical supersymmetric Lagrangian we write down the equations of motion which follow from (3.34) and (3.35). Writing L_{Scl} as

$$L_{Scl} = \frac{1}{2}\dot{x}^2 - \frac{1}{2}U'^2(x) + \frac{i}{2}\left(\psi\dot{\psi} - \dot{\psi}\psi\right) - U''(x)\psi\psi \qquad (3.36)$$

we see that the equations of motion are [see also (2.128)]

$$
\begin{aligned}
\dot{\psi} &= -iU''\psi \\
\dot{\overline{\psi}} &= iU''\overline{\psi} \\
\ddot{x} &= -(U'U'') - (U''')\overline{\psi}\psi \qquad (3.37)
\end{aligned}
$$

Setting $\psi_0 = \psi(0)$ and $\overline{\psi}_0 = \overline{\psi}(0)$ and looking for a solution of $x(t)$ of the type $x(t) = \overline{x}(t) + c(t)\overline{\psi}_0\psi_0$, the solutions for $\psi(t), \overline{\psi}(t)$ and $c(t)$ turn out to be

$$\psi(t) = \psi_0 \exp\left[-\int_0^t U''(\overline{x}(\tau))d\tau\right]$$

$$\overline{\psi}(t) = \overline{\psi}_0 \exp\left[i\int_0^t U''(\overline{x}(\tau))d\tau\right]$$

$$c(t) = \frac{\dot{\overline{x}}(t)}{\dot{\overline{x}}(0)}\left[c(0) + \frac{\dot{\overline{x}}(0)}{2}\int_0^t \frac{\lambda - U''(\overline{x}(\tau))}{\mu - \frac{1}{2}U'^2(\overline{x}(\tau))}d\tau\right] \quad (3.38)$$

In the expression for $c(t)$, λ and μ enter as arbitrary constants of integration but are linked to the conservation of energy $E = \mu + \lambda\overline{\psi}_0\psi_0$. On the other hand $\overline{x}(t)$ may be interpreted as the quasi-classical contribution to $x(t)$. A quasi-classical solution has the feature that it describes fully the classical dynamics of the bosonic along with fermionic degrees of freedom [13,14].

In conclusion let us note that spin is a purely quantum mechanical concept having no classical analogy. Thus we cannot think of constructing a classical wave packet having a spin $\frac{1}{2}$ angular momentum. Pseudo-classical mechanics is somewhat in between classical mechanics and quantum mechanics in that even in the limit $\hbar \to 0$ we can persist with Grassmann variables. Historically, the role of an anti-commuting variable in relation to the quantal action was explained by Schwinger [15]. Later, Matthews and Salam [16] tackled the problem of evaluating functional integrals over anticommuting functions. Berezin and Marinov [17] also developed the Grassmann variant of the Hamiltonian mechanics and in this way presented a generalization of classical mechanics.

3.4   References

[1] H. Goldstein, *Classical Mechanics*, Addison-Wesley, MA, 1950.

[2] E.C.G. Sudarshan and N. Mukunda, *Classical Dynamics: A Modern Perspective*, John Wiley & Sons, New York, 1974.

[3] M.G. Calkin, *Lagrangian and Hamiltonian Mechanics*, World Scientific, Singapore, 1996.

[4] N.D. Sengupta, *News Bull. Cal. Math. Soc.*, **10**, 12, 1987.

[5] R. Casalbuoni, *Nuovo Cim*, **A33**, 115, 389, 1976.

[6] J.L. Martin, *Proc. Roy. Soc.* **A251**, 536, 1959.

[7] L. Brink, S. Deser, B. Zumino, P. di Vecchia, and P. Howe, *Phys. Lett.*, **64B**, 435, 1976.

[8] A. Barducci, R. Casalbuoni, and L. Lusanna, *Nucl. Phys.*, **B124**, 93, 521, 1977.

[9] R. Marnelius, *Acta Phys. Pol.*, **B13**, 669, 1982.

[10] P.G.O. Freund, *Introduction to Supersymmetry*, Cambridge University Press, Cambridge, 1986.

[11] J. Barcelos - Neto, A. Das, and W. Scherer, *Phys. Rev.*, **D18**, 269, 1987.

[12] S.N. Biswas and S.K. Soni Pramana, *J. Phys.*, **27**, 117, 1986.

[13] G. Junker and S. Matthiesen, *J. Phys. A: Math. Gen.*, **27**, L751, 1994.

[14] G. Junker and S. Matthiesen, *J. Phys. A: Math. Gen.*, **28**, 1467, 1995.

[15] J. Schwinger, *Phil. Mag.*, **49**, 1171, 1953.

[16] P.T. Matthews and A. Salam, *Nuovo. Cim.*, **2**, 120, 1955.

[17] F.A. Berezin and M.S. Marinov, *Ann. Phys.*, **104**, 336, 1977.

CHAPTER 4

SUSY Breaking, Witten Index, and Index Condition

4.1 SUSY Breaking

As already noted in Chapter 2, while the Hamiltonian of the harmonic oscillator is invariant under the interchange of the lowering and raising operators, the vacuum, which is defined to be the lowest state, is not. On the other hand, when we speak of SUSY being an exact or an unbroken symmetry both the supersymmetric Hamiltonian H_s as well as its lowest state remain invariant with repsect to the interchange of the supercharge operators Q and Q^+. This is due to the cancellation (corresponding to $\omega = \omega_B = \omega_F$) between the bosonic and fermionic contributions to the ground state energy thus admitting of a zero-energy lowest state for the supersymmetric Hamiltonian.

Let us now study the case of SUSY being broken spontaneously [1,2]. We know from (2.47) that

$$
\begin{aligned}
E_0 &= <0|H_s|0> \\
&= |Q_1|0>|^2 > 0
\end{aligned}
\tag{4.1}
$$

where we do not assume a negative norm ghost state contributing.

57

So $Q_1|0 > \neq 0$ means existence of degenerate vacua related by the supercharge operator. This can be made more explicit by assuming specifically

$$Q|0 >= \lambda|0 >' \neq 0 \qquad (4.2)$$

where Q is defined according to (2.31). Since Q anti-commutes with H_s we have

$$
\begin{aligned}
H_s Q|0 > &= Q H_s|0 > \\
&= Q\mu|0 > \\
&= \lambda\mu|0 >'
\end{aligned}
\qquad (4.3)
$$

where μ is the ground-state eigenvalue of H_s. Also from (4.2) we can write

$$H_s Q|0 >= \lambda H_s|0 >' \qquad (4.4)$$

(4.3) and (4.4) thus point to

$$H_s|0 >'= \mu|0 >' \qquad (4.5)$$

showing $|0 >$ and $|0 >'$ to be degenerate states. The condition $E_0 > 0$ is as necessary as well as sufficient for SUSY to be spontaneously broken.

The previous steps can also be formulated in terms of the constraints on the functional forms of the superpotential. For unbroken SUSY we found from (2.56a) and (2.56b) that normalizability of the ground-state wave function requires $W(x)$ to differ in signs at $x \to \pm\infty$. This is of course the same as saying that $W(x)$ possesses an odd number of zeros in $(-\infty, \infty)$. However if $W(x)$ is an even function of x, there cannot be any normalizable zero-energy wave function and we are led to degenerate ground-states having a nonzero energy value. Such a situation corresponds to spontaneous supersymmetric breaking of SUSY. For example, if we take $W(x) = \frac{1}{2}gx^2$, g a coupling constant, the ground-state wave function behaves as $\exp\left(\pm\frac{1}{2}\int_{x_0}^x gx^2 dx\right)$ which obviously blows up either at plus or minus infinity.

Thus spontaneous breaking of SUSY is concerned with $E > 0$ with pairing of all energy levels. It is easy to see from (2.59b) and (2.59c) that the following interrelationships among the eigenfunctions of H_+ and H_- are implied

$$\left(W + \frac{d}{dx}\right)\psi_+^n = \sqrt{2E_n}\psi_-^n \tag{4.6a}$$

$$\left(W - \frac{d}{dx}\right)\psi_-^n = \sqrt{2E_n}\psi_+^n \tag{4.6b}$$

where $n = 1, 2, \ldots$ and the real-valued superpotential $W(x)$ is assumed to be continuously differentiable. The set (4.6) brings out the roles of H_+ and H_- namely

$$\left(-\frac{1}{2}\frac{d^2}{dx^2} + V_+\right)\psi_+^n = E_n\psi_+^n \tag{4.7a}$$

$$\left(-\frac{1}{2}\frac{d^2}{dx^2} + V_-\right)\psi_-^n = E_n\psi_-^n \tag{4.7b}$$

where $n = 1, 2, \ldots$ and (V_+, V_-) are given by (2.29).

4.2   Witten Index

To inquire into the nature of a system as to whether it is supersymmetric or spontaneously breaks SUSY, it is necessary to look for its zero-energy states. Consider the so-called Witten index [3-5] which is defined to count the difference between the number of bosonic and fermionic zero-energy states

$$\Delta \equiv n_B^{(E=0)} - n_F^{(E=0)} \tag{4.8}$$

This is logical since for energies which are strictly positive there is a pairing between the energy levels corresponding to the bosonic and fermionic states. Thus $\Delta \neq 0$ immediately signals SUSY to be unbroken as there does exist a state with $E = 0$.

For the spontaneously broken SUSY case note that the nonvanishing of classical potential energy implies the vacuum energy to be strictly positive in the classical approximation. A suitable example is $V(x) = \frac{1}{2}(x^2 + c)^2 \geq \frac{1}{2}c^2 > 0$ (for $c > 0$) and SUSY is spontaneously broken. On the other hand, if the vacuum energy is vanishing at the classical level, then SUSY prevails in perturbation theory and can be broken only through nonperturbative effects. However just from the vanishing of the quantity Δ it is not evident whether SUSY is spontaneouly broken $[n_B^{(E=0)} = 0 = n_F^{(E=0)}]$ or unbroken $[n_B^{(E=0)} = n_F^{(E=0)} \neq 0]$.

It is worth pointing out that as long as the basic supersymmetric algebra holds the various parameters, such as the mass or coupling constants, it can undergo changes leading to deformation in the potential. Such variations of parameters will, of course, also cause the energy of the states to change. But because of boson-fermion pairing in the supersymmetric theory the states must move (corresponding to their ascending or descending) in pairs. In other words Δ is invariant under the variation of parameters. This is the so-called "topologial invariance" of the Witten index. More specifically, the Witten index is insensitive to the variations of the parameters entering the potential so long as the asymptotic behaviour of $W(x)$ does not show any change in signs. For example, we can deform the function $W(x) = \lambda x(x^2 - c^2)$ by changing the parameters λ and c or even adding a quadratic term without affecting its zeros. However, if we add a quartic term we run into an undesirable situation where $W(x)$ has an extra zero. This causes a jump in Δ. Note that such deformations of $W(x)$ are disallowed.

The definition (4.8) for Δ suggests $Tr(-1)^{N_f}$ to be a natural representation for it

$$\begin{aligned} \Delta &= Tr(-1)^{N_f} \\ &= Tr(1 - 2N_f) \end{aligned} \tag{4.9}$$

where N_f is the fermion number operator. It is clear that $(-1)^{N_f}$ assumes the value $+1$ or -1 accordingly as there are even or odd number of fermions.

However, the above definition of the trace in terms of the fermion number operator needs regularization. This is because the trace is taken over the Hilbert space and issues relating to convergence may arise. In the following we must first look into the anomalous behaviour of Δ in a finite temperature theory when (4.9) is used naively. Afterward we take up the regularization of Δ.

4.3   Finite Temperature SUSY

An obvious place to expect [7,8] breaking of SUSY is in finite temperature domains where the thermal distribution of bosons and fermions are different. The connection of Δ to β ($\equiv \frac{1}{kT}$, k the Boltzmann constant) gives a clue to SUSY breaking one expects Δ, starting from a nonzero value at zero-temperature, to vanish at finite temperature.

In the literature the subject of thermofield dynamics [9] has proved to be an appropriate formulation of the thermal quantum theory. As has been widely recognized, the two-mode squeezed state is some kind of a thermofield state whose evolution can be described by the Wigner function [10]. A novel aspect of two-mode squeezing [11,12] is the creation of thermal-like noise in a pure state. The problem lacks somewhat in uniqueness since there arise inevitable ambiguities in the precise identification of the relevant operators performing squeezing. Neverthelss, it is important to bear in mind that squeezing is essentially controlled by a generator that is bilinear in bosonic variables and that all the essential features of squeezing are present in the state obtained by operating the generator on the vacuum.

It is important to realize that the two-mode squeezed state can be associated with the following basic commutation relations satisfied by the coordinates and momenta

$$[x_1, p_1] = i, \quad [x_2, p_2] = -i \tag{4.10}$$

In terms of the oscillators $b_i (i = 1, 2)$, the pairs (x_1, p_1) and (x_2, p_2) are

$$x_1 = \frac{1}{\sqrt{2}} \left(b_1 + b_1^+ \right), \quad p_1 = \frac{1}{i\sqrt{2}} \left(b_1 - b_1^+ \right) \tag{4.11a}$$

$$x_2 = \frac{1}{\sqrt{2}} \left(b_2 + b_2^+ \right), \quad p_1 = -\frac{1}{i\sqrt{2}} \left(b_2 - b_2^+ \right) \tag{4.11b}$$

That the two quantum conditions in (4.10) need to differ in sign arises from the necessity to preserve the so-called Bogoliubov transformation

$$b_1(\beta) = b_1 \cosh \theta(\beta) - b_2^+ \sinh \theta(\beta) \tag{4.12a}$$

$$b_2(\beta) = b_2 \cosh \theta(\beta) - b_1^+ \sinh \theta(\beta) \tag{4.12b}$$

Note that $(4.12a, b)$ have been obtained from the transformations

$$
\begin{aligned}
b_1(\beta) &= U(\beta) b_1 U^{-1}(\beta) \\
b_2(\beta) &= U(\beta) b_2 U^{-1}(\beta) \\
U(\beta) &= \exp\left[-\theta(\beta)(b_1 b_2 - b_2^+ b_1^+)\right]
\end{aligned}
\tag{4.13}
$$

The two-mode squeezed Bogoliubov transformation is also often referred to as the thermal Bogoliubov transformation. Note that the sets (b_1, b_1^+) and (b_2, b_2^+) continue to satisfy the normal bosonic commutation relations

$$
\begin{aligned}
\left[b_1, b_1^+\right] &= 1 \\
\left[b_2, b_2^+\right] &= 1 \\
[b_1, b_2] &= 0 \\
\left[b_1^+, b_2\right] &= 0
\end{aligned}
\tag{4.14}
$$

Denoting the vacuum of the system (b_1, b_2) by $|0>$ and that of $(b_1(\beta), b_2(\beta))$ by $|0(\beta)>$ it follows that

$$
\left|0(\beta)>= e^{-\ln \cosh \theta(\beta)} e^{-b_1^+ b_2^+ \tanh \theta(\beta)}\right|0>
\tag{4.15}
$$

with $b_i|0>= 0$, $b_i(\beta)|0(\beta)>= 0$, $i = 1$ and 2. It is clear from the above that the $b_1 b_2$ pairs are condensed.

With this brief background on two-mode squeezing, let us define the thermal annihilation operators $a_i, i = 1$ and 2 for fermions namely

$$
\begin{aligned}
a_1(\beta) &= a_1 \cos \theta(\beta) - a_2^+ \sin \theta(\beta) \\
a_2(\beta) &= a_2 \cos \theta(\beta) + a_1^+ \sin \theta(\beta)
\end{aligned}
\tag{4.16}
$$

Analogous to (4.15), one can write down the thermal vacuum for fermionic oscillators which consists of $a_1 a_2$ pairs.

The boson-fermion manifestation in a supersymmetric theory suggests that the underlying thermal vacuum is given by

$$
\left|0(\beta)>= \exp\left[-\theta(\beta)(a_1 a_2 - a_2^+ a_1^+) - \theta(\beta)(b_1 b_2 - b_2^+ b_1^+)\right]\right|0>
\tag{4.17}
$$

where $|0>$ is the vacuum at $T = 0$ and $\tan \theta(\beta) = e^{-\beta/2}$.

If one now uses the definition (4.9) for Δ corresponding to $N_F = a_1^+ a_1$ in thermal vacuum, one finds [7] from (4.16)

$$
\begin{aligned}
\Delta &= \; <0(\beta)|(1 - 2a_1^+ a_1)|0(\beta)> \\
&= \frac{1 - e^{-\beta}}{1 + e^{\beta}}
\end{aligned}
\tag{4.18}
$$

It transpires from (4.18) that as $T \to 0$, the index $\Delta \to 1$ while as $T \to \infty, \Delta \to 0$. However for any intermediate value of T in the range $(0, \infty), \Delta$ emerges fractional and so the definition (4.9) is not a good representation for the index.

We now look into a heat kernel regularized index. We wish to point out that even for such a regularized index the β dependence persists when one considers the presence of a continuum distribution.

4.4 Regulated Witten Index

There have been several works on the necessity of a properly regularized Witten index. Witten himself proposed [3]

$$
\Delta_\beta = Tr\left[(-1)^{N_F} e^{-\beta H}\right]
\tag{4.19}
$$

while Cecotti and Girardello considered [13-21] a functional integral for Δ_β

$$
\Delta_\beta = \int [d\Phi] \, e^{-S_\beta(\Phi)}
\tag{4.20}
$$

where the measure $[d\Phi]$ runs over all field configurations satisfying periodic boundary conditions and S the Euclidean action. It has been found that one can evaluate Δ_β both with and without the use of constant configurations [22]. Other forms of a regularized index have also been adopted in the literature. See, for example, [13-21].

The regularized Witten index Δ_β is, in general, β-dependent when the theory contains a continuum distribution apart from discrete states [6]. This is in contrast to our normal expectations that since $E \neq 0$ states do not contribute to the trace, the right hand side of (4.19) should be independent of β. Of course Δ_β is independent of β if the Hamiltonian shows discrete spectrum. In the following let us study the β-dependence of Δ_β.

Expressing (4.19) as

$$\Delta_\beta = Tr\left(e^{-\beta H_+} - e^{-\beta H_-}\right) \tag{4.21}$$

and defining the kernels corresponding to H_+ and H_- to be

$$K_\pm(x, y, \beta) = < y|e^{-\beta H_\pm}|x > \tag{4.22}$$

we can write Δ_β as

$$\Delta_\beta = \int dx \left[K_+(x, x, \beta) - K_-(x, x, \beta)\right] \tag{4.23}$$

It is also implied from (4.22) that

$$K_\pm(x, y, \beta) = \sum_k e^{-\beta E_k} \psi_k^\pm(x)\psi_k^\pm(y) \tag{4.24}$$

in which the contributions from the discrete and continuum states can be separated out explicitly as

$$K_\pm(x, x, \beta) = \sum_k e^{-\beta E_k} \psi_k^\pm(x)\psi_k^\pm(x) + \int dE e^{-\beta E_k}\psi_E^\pm(x)\psi_E^\pm(x) \tag{4.25}$$

It is useful to distinguish the continuum state by $\psi(k, x)$. From (4.7) we can deduce $E(k) = \frac{1}{2}k^2 + W_0^2 > W_0^2$ corresponding to $W(x) \to \pm W_0$ for $x \to \pm\infty$ noting that one can construct solutions of (4.7) of the types $e^{\pm ikx}$ as $x \to \pm\infty$, respectively.

To evaluate the integrand of (4.23) we must first take help from (4.24) to express

$$-\frac{d}{d\beta}[K_+(x, y, \beta) - K_-(x, y, \beta)] = \sum_k e^{-\beta E(k)}\Psi(x, y, k) \tag{4.26}$$

where $\sum_k$ accounts for the bound states as well as continuum contributions and the quantity $\Psi(x, y, k)$ stands for

$$\Psi(x, y, k) = \psi_+^*(y, k)\psi_+(x, k) - \psi_-^*(y, k)\psi_-(x, k) \tag{4.27}$$

Using the supersymmetric equations (4.7) we now obtain

$$E(k)\Psi(x,y,k)$$

$$
\begin{aligned}
= &\ \frac{1}{2}\left[\frac{1}{2}\left\{-\frac{d^2}{dy^2}+W^2(y)-W'(y)\right\}\psi_+^*(y,k)\psi_+(x,k)\right.\\
&+\frac{1}{2}\left\{-\frac{d^2}{dx^2}+W^2(x)-W'(x)\right\}\psi_+^*(y,k)\psi_+(x,k)\\
&\left.-\left\{\frac{d}{dy}+W(y)\right\}\psi_+^*(y,k)\left\{\frac{d}{dx}+W(x)\right\}\psi_+(x,k)\right]\\
= &\ \frac{1}{2}\left[\frac{1}{2}\left\{-\psi_+(x,k)\frac{d^2}{dy^2}\psi_+^*(y,k)-\psi_+^*(y,k)\frac{d^2}{dx^2}\psi_+(x,k)\right\}\right.\\
&+\frac{1}{2}\psi_+(x,k)\left\{W^2(y)-W'(y)\right\}\psi_+^*(y,k)\\
&+\frac{1}{2}\psi_+^*(y,k)\left\{W^2(x)-W'(x)\right\}\psi_+(x,k)\\
&-\left\{\frac{d\psi_+^*(y,k)}{dy}+W(y)\psi_+^*(y,k)\right\}\\
&\left.\left\{\frac{d\psi_+(x,k)}{dx}+W(x)\psi_+(x,k)\right\}\right]
\end{aligned}
\tag{4.28}
$$

Putting $x=y$ it follows that

$$
\begin{aligned}
E(k)\Psi(x,x,k)\ =&\ \frac{1}{2}\left[-\frac{1}{2}\left\{\psi_+(\psi_+^*)''+\psi_+^*\psi_+''\right\}\right.\\
&+\frac{1}{2}\psi_+(W^2-W')\psi_+^*+\frac{1}{2}\psi_+^*(W^2-W')\psi_+\\
&\left.-\left\{(\psi_+^*)'+W\psi_+^*\right\}\left\{\psi_+'+W\psi_+\right\}\right]
\end{aligned}
\tag{4.29}
$$

where the dependence on (x,k) of ψ_+,ψ_+^* and W has been suppressed. Simplifying the right hand side one finds

$$
\begin{aligned}
E(k)\Psi(x,x,k)\ =&\ -\frac{1}{4}\left[\psi_+^*\psi_+''+\psi_+(\psi_+^*)''+2(\psi_+^*)'\psi_+'\right.\\
&\left.+2\frac{d}{dx}(W\psi_+\psi_+^*)\right]\\
=&\ -\frac{1}{4}\frac{d}{dx}\left[\frac{d}{dx}(\psi_+^*\psi_+)+2W\psi_+^*\psi_+\right]\\
=&\ -\frac{1}{4}\frac{d}{dx}\left(\frac{d}{dx}+2W\right)(\psi_+^*\psi_+)
\end{aligned}
\tag{4.30}
$$

From (4.26) one thus has [6]

$$\frac{d}{d\beta}\left[K_+(x,x,\beta) - K_-(x,x,\beta)\right] = \frac{1}{4}\frac{d}{dx}\left(\frac{d}{dx} + 2W\right)K_+(x,x,\beta)$$

(4.31)

The above identity greatly facilitates the computation of $\frac{d\Delta_\beta}{d\beta}$. Indeed inserting (4.31) in the right hand side of (4.23) we can project out the β-dependence of Δ_β in a manner

$$\frac{d\Delta_\beta}{d\beta} = \frac{1}{4}\int dx\frac{d}{dx}\left(\frac{d}{dx} + 2W\right)K_+(x,x,\beta)$$

(4.32)

which can also be expressed, using (4.24), and (4.6a) as

$$\frac{d\Delta_\beta}{d\beta} = \frac{1}{2}\int dx\frac{d}{dx}\sum_{k\neq 0}e^{-\beta E_k}\sqrt{2E_k}\psi_k^+\psi_k^-$$

(4.33)

Let us consider now the solitomic example $W(x) = \tanh x$ corresponding to which the supersymmetric partner Hamiltonians are

$$H_+ = -\frac{1}{2}\frac{d^2}{dx^2} + \frac{1}{2}\left(1 - 2\sec h^2 x\right)$$

(4.34a)

$$H_- = -\frac{1}{2}\frac{d^2}{dx^2} + \frac{1}{2}$$

(4.34b)

If one employs periodic boundary conditions over the internal $[-L, L]$ the associated wave functions of $H_\pm$ turn out to be of the following two types

$$\psi_-^1 = N\ \cos k_1 x$$

(4.35a)

$$\psi_+^1 = \frac{N}{\sqrt{k_1^2 + 1}}\{k_1\sin k_1 x + \tanh x \cos k_1 x\}$$

(4.35b)

$$\psi_-^2 = N\ \sin k_2 x$$

(4.36a)

$$\psi_+^2 = \frac{N}{\sqrt{k_2^2 + 1}}\{-k_2\cos k_2 x + \tanh x \sin k_2 x\}$$

(4.36b)

These wave functions are subjected to

$$k_1 > 0:\ k_1\sin k_1 L + \tanh L \cos k_1 L\ =\ 0$$

(4.37)

$$k_2 > 0:\ \sin k_2 L\ =\ 0$$

(4.38)

which have been obtained from the considerations $\psi_\pm^i(L) = \psi_\pm^i(-L)$, $i = 1, 2$ and $E(k_i) = \frac{1}{2}(1 + k_i^2), i = 1, 2$. The energy constraints from the Schroedinger equations $H_+\psi_+^i = E(k_i)\psi_+^i, i = 1$ and 2. It is worth mentioning that (4.35) and (4.36) are consistent with the intertwining conditions (4.6).

We can now use the formula (4.33) to calculate $\frac{d\Delta_\beta}{d\beta}$ corresponding to the wave functions (4.35) and (4.36). It is trivial to check that as a result of the associated boundary conditions $\frac{d\Delta_\beta}{d\beta} = 0$. [Note that in addition to (4.35b) and (4.36b) H_+ also possesses a normalizable zero-energy state $\lambda\mathrm{sech}^2x$, λ a constant, but it does not contribute to the sum in (4.33)].

We now pass on the limit $L \to \infty$ when one recovers the continuum states. Employing the usual normalization $N \to \frac{1}{\sqrt{\pi}}$, the derivative $\frac{d\Delta_\beta}{d\beta}$ is found to be

$$
\begin{aligned}
\frac{d\Delta_\beta}{d\beta} &= \frac{1}{2\pi}\int_0^\infty dk \frac{e^{-\frac{\beta}{2}(k^2+1)}}{\sqrt{k^2+1}}\left[\cos kx\left\{k\sin kx + \tanh x\cos kx\right\}\right. \\
&\qquad \sqrt{1+k^2} + \sin kx\left\{-k\cos kx + \tanh x\sin kx\right\} \\
&\qquad \left.\sqrt{1+k^2}\right]_{x=-\infty}^{x=+\infty} \\
&= \frac{1}{2}\frac{e^{-\frac{\beta}{2}}}{2\pi}\left\{\int_0^\infty e^{-\frac{\beta k^2}{2}}\,dk\right\}[\tanh x]_{x-\infty}^{x=+\infty} \\
&= \frac{1}{2}\frac{e^{-\frac{\beta}{2}}}{\sqrt{2\pi\beta}} \tag{4.39}
\end{aligned}
$$

which, clearly, is β-dependent.

The above discussions give us an idea on the behaviour of a regulated Witten index. Actually the evaluation of the index depends a great deal on the choice of the method adopted and in finding a suitable regularization procedure. Also the behaviour of Δ_β depends much on the nature of the spectrum; a purely continuous one extending to zero may yield fractional values of Δ_β. In this connection note that if we use the representation (4.32) we run into the problem of determining exactly the heat kernels. There is also the related issue of the viability of the interchange of the k and x limits of integration. For more on the anomalous behaviour of the Witten index and its judicious computation using the heat kernel techniques one may consult [6].

4.5   Index Condition

We now analyze the Fredholm index of the annihilation operator b defined by [23,24]

$$\delta \equiv \dim ker\, b - \dim ker\, b^+ \tag{4.40}$$

where b and b^+ are, respectively, the annihilation and creation operators of the oscillator algebra given by (2.6) and (2.7). In (4.40) dim ker corresponds to the dimension of the space spanned by the linearly independent zero-modes of the relevant operator. Since $b|0> = 0$ it is obvious that dim ker $b = \{|0>\}$ while dim ker b^+ is empty. Thus $\delta = 1$.

Apart from (4.40) we also have [24]

$$\dim ker\, b^+ b - \dim ker\, bb^+ = 1 \tag{4.41a}$$

where dim ker $b^+ b$ represents the number of normalizable eigenkets $|\psi_n>$ obeying $b^+ b|\psi_n> = 0$. To avoid a singular point in the index relation it is useful to restrict dim ker $b^+ b < \infty$. A deformed quantum condition often leads to the existence of multiple vacua when δ may become ill-defined.

It is interesting to observe that for a truncated oscillator one finds in place of (4.41a)

$$\dim ker\, b_s^+ b_s - \dim ker\, b_s b_s^+ = 0 \tag{4.41b}$$

where b_s and b_s^+ are the truncated vertions [25] of b and b^+ defined in an $(s+1)$-dimensional Fock space. To establish (4.41b) we transform the eigenvalue equations

$$b_s^+ b_s \phi_n = e_n^2 \phi_n \tag{4.42a}$$

to the form

$$b_s b_s^+ \chi_n = e_n^2 \chi_n \tag{4.42b}$$

by setting $\chi_n = \frac{1}{e_n} b_s \phi_n$ with $e_n \neq 0$. We thus find the normalizability of the eigenfunctions ϕ_n and χ_n to go together. For a finite-dimensional matrix representation we also have $Tr(b_s^+ b_s) = Tr(b_s b_s^+)$. In this way we are led to (4.41b).

When applied to SUSYQM δ can be related to the Witten index which counts the difference between the number of bosonic end fermionic zero-energy states. This becomes transparent if we focus on (4.41a). Replacing the bosonic operators (b, b^+) by (A, A^+) according to (2.34) in terms of the superpotential $W(x)$, the left hand side of (4.41a) just expresses the difference between the number of bosonic and fermionic zero eigenvalues. We thus have a correspondence with (4.8).

In the following we study [26] the Fredholm index condition (4.40) for the annihilation operator of the deformed harmonic oscillator and *q*-parabose systems in a generalized sense to show how multiple vacua may arise. We also look into the singular aspect of δ and point out some remedial measures to have an ambiguous interpretation of δ.

4.6 *q*-deformation and Index Condition

Interest in quantum deformation seems to have started after the work of Kuryshkin [27-30] who considered a *q*-deformation in the form $\mathcal{A}\mathcal{A}^+ - q\mathcal{A}^+\mathcal{A} = 1$ for a pair of mutually adjoint operators $\mathcal{A}$ and $\mathcal{A}^+$ to study interactions among various particles. Later Janussis et al. [31], Biedenharn [32], Macfarlane [33], Sun and Fu [34] and several others [35-48] made a thorough analysis of deformed structures with a view to inquiring into plausible modifications of conventional quantum mechanical laws. Recent interest in quantum deformation comes from its link [49,50] with anyons and Chern-Simmons theories. The ideas of *q*-deformation has also been intensely pursued to develop enveloping algebras [51] quasi-Coherent states [52], rational conformal field theories [53] and geometries possessing non commutative features [54].

Let us consider the following standard description of a *q*-deformed harmonic oscillator

$$\mathcal{A}\mathcal{A}^+ - q\mathcal{A}^+\mathcal{A} = q^{-N}, \ q \in (-1, 1) \tag{4.42}$$

with $\mathcal{A}$ and $\mathcal{A}^+$ obeying

$$[N, \mathcal{A}] = -\mathcal{A}, \ [N, \mathcal{A}^+] = \mathcal{A}^+ \tag{4.43}$$

As the deformation parameter $q \to 1, \mathcal{A} \to b$ and we recover the familiar bosonic condition $bb^+ - b^+b = 1$ for the normal harmonic oscillator.

The operators $(\mathcal{A}, \mathcal{A}^+)$ may be related to the bosonic annihilation and creation operators b and b^+ by writing

$$\mathcal{A} = \Phi(N)b, \quad \mathcal{A}^+ = b^+\Phi(N) \tag{4.44}$$

where N is the number operator b^+b and Φ any function of it.

Exploiting the eigenvalue equation $\Phi(N_B)|n>== \Phi(n)|n>$, the representations (4.44) lead to the recurrence relation

$$(n+1)\Phi^2(n) - qn\Phi^2(n-1) = q^{-n} \tag{4.45}$$

The above equation has the solution

$$\Phi(n) = \sqrt{\frac{1}{n+1}\frac{q^n - q^{-n}}{q - q^{-1}}} \tag{4.46}$$

which implies that $\Phi(N)$ is given by

$$\Phi(N) = \sqrt{\frac{[N+1]}{N+1}} \tag{4.47}$$

with

$$[x] = \frac{q^x - q^{-x}}{q - q^{-1}} \tag{4.48}$$

From (4.44) and (4.48) the Hamiltonian for the q-deformed harmonic oscillator can be expressed as

$$\begin{aligned} H^d &= \frac{\omega}{2}\{\mathcal{A}, \mathcal{A}^+\} \\ &= \frac{\omega}{2}\{[N+1] + [N]\} \end{aligned} \tag{4.49}$$

Note that the commutator $[\mathcal{A}, \mathcal{A}^+\}$ reads

$$[\mathcal{A}, \mathcal{A}^+] = [N+1] - [N] \tag{4.50}$$

Just as we worked out the supersymmetric Hamiltonian in Chapter 2 as arising from the superposition of bosonic and fermionic oscillators, here too we can think of a q-deformed SUSY scheme [55-59] by considering q-superoscillators. Indeed we can write down a q-deformed supersymmetric Hamiltonian H_s^q in the form

$$H_s^q = \frac{\omega}{2}\left(\{\mathcal{A}, \mathcal{A}^+\} + [F^+, F]\right) \tag{4.51}$$

with $\mathcal{A}$ and $\mathcal{A}^+$ obeying (4.42) and (4.43) and F, F^+ are, respectively, the q-deformed fermionic annihilation and creation operators subjected to

$$FF^+ + qF^+F = q^{-M} \qquad (4.52)$$

In contrast to the usual fermionic operators whose properties are summarized in (2.12) - (2.14), the deformed operators F and F^+ do not obey the nilpotency conditions: $(F)^n \neq 0$, $(F^+)^n \neq 0$ for $n > 1$ and $q \in (0,1)$. This means that any number of q-fermions can be present in a given state. For a study of the properties of q-superoscillators in SUSYQM see [55].

To evaluate δ we need to look into the plausible ground states of H^d (note that H^d forms a part of H^q_s). As such we have to search for those states which are annihilated by the deformed operator $\mathcal{A}$.

In the following we shall analyze Ferdholm index δ for those situations when the elements in dim ker $\mathcal{A}$ as well as dim ker $\mathcal{A}^+$ are countably infinite and so the index, when evaluated naively, may not be a well-defined quantity. Indeed such a possibility occurs when the Fock space of the underlying physical system is deformed and the deformation parameter is assumed complex with modules unity (for preservation of hermiticity)

$$q = e^{\pm \frac{2i\pi}{k+1}}, \ |q| = 1, \ k > 1 \qquad (4.53)$$

However we shall argue that since both the kernels (coresponding to $\mathcal{A}$ and $\mathcal{A}^+$) turn out to be countably infinite modulo $(k+1)$, the question of building up an infinite sequence of eigenkets (on repeated application to the ground state by the creation operator) is ruled out and the deformed system has to choose its ground state along with the spectrum over some suitable finite dimensional Fock space. This has the consequence of transfering δ from the ill-defined $(\infty - \infty)$ to the zero-value.

For the deformed oscillator the roles of $\mathcal{A}$ and $\mathcal{A}^+$ are

$$\mathcal{A}|k> = \sqrt{[k]}|k-1> \qquad (4.54a)$$

$$\mathcal{A}^+|k> = \sqrt{[k+1]}|k+1> \qquad (4.54b)$$

where

$$|k> = \frac{1}{\sqrt{[k]!}} (\mathcal{A}^+)^k |0> \tag{4.55}$$

We see that the bracket $[k+1] = 0$ whenever q assumes values (4.53) for which $\Phi(k) = 0$ (We do not consider irrational values of θ in the present context). It follows that for these values of q the Fock space gets split into finite-dimensional sub-spaces $\mathcal{T}_k$. One can thus think of the kernel $\mathcal{A}$ to consist of a countably infinite number of elements starting from $|0>$ with the subsequent zero-mades placed at $(k+1)$-distance from each other. Similar reasoning also holds for the kernel of $\mathcal{A}^+$. We can write

$$ker\ \mathcal{A} = \{|0>, |k+1>, |2k+2>, \ldots\} \tag{4.56a}$$

$$ker\ \mathcal{A}^+ = \{|k>, |2k+1>, \ldots\} \tag{4.56b}$$

Note that since ker $\mathcal{A}^+$ is nonempty, the process of creating higher states by repeated application of $\mathcal{A}^+$, on some chosen vacuum belonging to a particular $\mathcal{T}_k$, has to terminate. By simple counting (which we illustrate in the more general parabose case below) δ corresponding to (4.56) takes the value zero.

Fujikawa, Kwek, and Oh [60] have shown that for values of q corresponding to (4.53) the singular situation discussed above allows for a hermitean phase operator as well as a nonhermitean one. They have argued that since rational values of θ are densely distributed over $\theta \in \mathbb{R}$, the notion of continuous deformation for the index cannot be formally defined which means, in consequence, that singular points associated with a rational θ are to be encountered almost everywhere. These authors have also shown how to avoid the problem of negative norms for $q = e^{2\pi i\theta}$.

4.7   Parabosons

The particle operators c and c^+ of a parabose oscillator obey the trilinear commutation relation [61,62]

$$[c, H] = c \tag{4.57}$$

where H is the Hamiltonian

$$H = \frac{1}{2} (cc^+ + c^+c) \tag{4.58}$$

The spectrum of states of parabosons of order p can be deduced by defining a shifted number operator

$$N = H - \frac{p}{2} \tag{4.59}$$

and postulating the existence of a unique vacuum which is subject to

$$c|0 > \ = \ 0 \tag{4.60}$$
$$N|0 > \ = \ 0 \tag{4.61}$$

While (4.59) implies that the commutation relations

$$[N, c] = -c, \quad [N, c^+] = c^+ \tag{4.62}$$

hold so that c and c^+ may be interpreted, respectively, as the annihilation and creation operator for the parabose states, the conditions (4.60) and (4.61) prescribe an additional restriction on $|0 >$ from (4.59)

$$c\,c^+|0 >= \ p|0 > \tag{4.63}$$

Using (4.63) one can not only construct the one-particle state $|1 >= \frac{c^+}{\sqrt{p}}|0 >$, $p \neq 0$ but also recursively the n-particle state.

However, one has to distinguish between the even and odd nature of the states as far as the action of c and c^+ on them is concerned

$$c|2n > \ = \ \sqrt{2n}|2n - 1 > \tag{4.64}$$
$$c|2n + 1 > \ = \ \sqrt{2n + p}|2n > \tag{4.65}$$
$$c^+|2n > \ = \ \sqrt{2n + p}|2n + 1 > \tag{4.66}$$
$$c^+|2n + 1 > \ = \ \sqrt{2n + 2}|2n + 2 > \tag{4.67}$$

It is obvious from the relations (4.64) - (4.67) that these reduce to the harmonic oscillator case when $p = 1$. It is also worth pointing out that for zero-order parabosons the above properties allow for the possibilty of nonunique ground states [61]. Indeed one readily finds from (4.64) - (4.67) that in addition to $c|0 >= 0$, the state $|0 >$ also obeys the relation $c^+|0 >= 0$. Also $c|1 >= 0$. However, for a physical system, though the level $|1 >$ is of higher energy, owing to the condition $c|1 >= 0$ the operator c can not connect $|1 >$ to level

$|0>$. As such $|0>$ plays the role of a spectator leaving $|1>$ as the logical choice of the ground state.

From the point of view of the Fredholm index defined for the parabox system as

$$\delta_p \equiv \dim \ker c - \dim \ker c^+ \tag{4.68}$$

(where obviously dim ker α corresponds to the dimension of the space spanned by the linearly independent zero-modes of the operator α), the elements of both dim ker c and dim ker c^+ for the $p = 0$ case are nonempty and finite

$$
\begin{aligned}
\dim \ker c &= \{|0>, |1>\} & (4.69) \\
\dim \ker c^+ &= \{|0>\} & (4.70)
\end{aligned}
$$

Taking the difference the index δ_p turns out to be 1. (Acutally for the harmonic oscillator as well as for the normal parabosons, $|0>$ is the true and genuine vacuum so that the unity value of the index holds trivially). It is also interresting to observe that even the physical interpretation offered above, which distinguishes $|1>$ as the natural choice for the ground state of $p = 0$ parabosons, also leads to $\delta_p = 1$.

We now turn to the case of deformed parabose oscillator of order $[p]$.

4.8   Deformed Parabose States and Index Condition

Let us carry out deformation of the parabose oscillators by replacing the eigenvalues in (4.64) - (4.67) by their q-brackets [62,63].

$$
\begin{aligned}
B|2n> &= \sqrt{[2n]}|2n-1> & (4.71) \\
B|2n+1> &= \sqrt{[2n+p]}|2n> & (4.72) \\
B^+|2n> &= \sqrt{[2n+p]}|2n+1> & (4.73) \\
B^+|2n+1> &= \sqrt{[2n+2]}|2n+2> & (4.74)
\end{aligned}
$$

As in the case of deformed harmonic oscillators here too we can connnect the operators (B, B^+) to the bosonic ones (b, b^+) through

$$B = \Phi_p(N)b, \quad B^+ = b^+ \Phi_p^+(N) \tag{4.75}$$

where

$$\Phi_p(N) = \sqrt{\frac{[N+p]}{N+1}}, \quad N \text{ even} \tag{4.76}$$

$$= \sqrt{\frac{[N+1]}{N+1}}, \quad N \text{ odd} \tag{4.77}$$

Accordingly we get for the Hamiltonian of the q-deformed para-bosons the expressions

$$H_{2n}^{pd} = \frac{1}{2}\{[2N+p]+[2N]\} \tag{4.78}$$

$$H_{2n+1}^{pd} = \frac{1}{2}\{[2N+p]+[2N+2]\} \tag{4.79}$$

where the suffixes indicate that H^{pd} is to operate on these states.

The transformations (4.75) mean that we are adopting a de-formed quantum condition of the form

$$BB^+ - q^\lambda B^+ B = f(q, N, [p]) \tag{4.80}$$

where

$$\lambda = p \text{ or } 2 - p$$
$$f(q, N, [p]) = q^{-2N}[p] \text{ or } q^{-2N-p}[2-p] \tag{4.81}$$

depending on whether (4.80) operates on the ket $|2n>$ or $|2n+1>$. We may remark that when $p = 1$, the q-deformed relation (4.80) reduces to (4.42) which is as it should be. Further hermiticity is preserved in (4.80) for all real q and also for complex values if q is confined to the unit circle $|q| = 1$.

To examine the index condition for the q-parabose system con-trolled by the Hamiltonians (4.78) and (4.79) we note that B and B^+ can be expanded as

$$B = \sum_{k=0}^{\infty} \sqrt{[2k+p]}|2k><2k+1|$$
$$+ \sum_{k=0}^{\infty} \sqrt{[2k]}|2k-1><2k| \tag{4.82}$$

$$B^+ = \sum_{k=0}^{\infty} \sqrt{[2k+p]}|2k+1><2k|$$

$$+ \sum_{k=0}^{\infty} \sqrt{[2k+2]}|2k+2><2k+1| \qquad (4.83)$$

Taking $q = \exp(2\pi i\theta)$ and setting $\theta = \frac{1}{2k+1}$, $k > \frac{1}{2}$, which implies $[2k+1] = 0$, it is clear (we consider rational values of θ only) from the first term in the right hand side of (4.82) that unless $p = 1$ the coefficient of $|2k><2k+1|$ remains nonvanishing except for cases when p assumes certain specific values constrained by the relation $p = 1 + 2m + m'$, where $k = m/m'$, m and m' are integers and $m' \neq 0$. However, barring these values of p (and of course $p = 1$), there is no other suitable choice of q for which this coefficient can be made zero. Similarly, for B^+. We therefore conclude that, for such a scenario, ker $B = \{|0>\}$, ker $B^+ = $ empty implying that the condition $\delta = 1$ holds.

On the other hand if $\theta = \frac{1}{2k}$ it results in the possibility of a singular situation with dim ker $B = \infty$ and dim ker $B^+ = \infty$

$$\text{ker } B = \{|0>, |2k>, |4k>, \ldots\} \qquad (4.84a)$$

$$\text{ker } B^+ = \{|2k-1>, |4k-1>, \ldots\} \qquad (4.84b)$$

We would like to stress that the above kernels depict a case similar to the truncated oscillator problem where the available degrees of freedom are finite [25]. In the present setup the degrees of freedom are those from the chosen vacuum state to the one just before the next member in the kernel.

Let us suppose, for concreteness, that $|2k>$ is the lowest state, then from (4.84a) the plausible states over which the system can run are those from $|2k>$ to $|4k-1>$. These are only finite in number. The point is that once a system takes a particular k-value (including $|0>$) for its ground state, all other states in the kernel and the accompanying higher states (which can be created from them) are rendered isolated in the sense that these are disjoint from the ones constructed by starting from a different vacuum. It may be noted that corresponding to the choice $|2k>$ made for the vacuum, the lowest k-value for the ground state in ker B^+ can be $|2k>$ since the state $|2k-1>$ in (4.84b) is lower to $|2k>$ and so acts like a spectator state as in the $p = 0$ case discussed earlier. It follows that not only for the kernels (4.84) but also for (4.56) the index vanishes.

4.9 Witten's Index and Higher-Derivative SUSY

In Witten's model of SUSYQM the operators A and A^+ are assumed to have first-derivative representations. However one can look for extensions of SUSYQM by resorting to higher-derivative versions of A and A^+. Apart from being mathematically interesting, these models of higher derivative SUSY (HSUSY) offer the scope of connections to nontrivial quantum mechanical systems as have been found out recently [64-78].

First of all we note that the factorization of H_s carried out in (2.36) is also consistent with the following behaviour of $H_\pm$ vis-a-vis the operators A and A^+

$$
\begin{aligned}
H_+ A^+ &= A^+ H_- \\
AH_+ &= H_- A
\end{aligned}
\tag{4.85}
$$

These interesting relations also speak of the double degeneracy of the spectrum with the ground state ψ_0^+, associated with the H_+ component, being nondegenerate.

As a first step towards building HSUSY, we assume that the underlying interwinning operators $\mathcal{A}$ and $\mathcal{A}^+$ are given by second-order differential representations

$$
\begin{aligned}
\mathcal{A} &= \frac{1}{2}\frac{d^2}{dx^2} + 2p(x)\frac{d}{dx} + g(x) \\
\mathcal{A}^+ &= \frac{1}{2}\frac{d^2}{dx^2} - 2p(x)\frac{d}{dx} + g(x) - 2p'(x)
\end{aligned}
\tag{4.86}
$$

where $p(x)$ and $g(x)$ are arbitrary but real functions of x. The above forms of $\mathcal{A}$ and $\mathcal{A}^+$ generate, as we shall presently see, the minimal version of HSUSY namely the second-derivative supersymmetric (SSUSY) scheme.

In terms of $\mathcal{A}$ and $\mathcal{A}^+$ we associate the corresponding super-charges q and q^+ as

$$
q = \begin{pmatrix} 0 & \mathcal{A} \\ 0 & 0 \end{pmatrix}, \quad q^+ = \begin{pmatrix} 0 & 0 \\ \mathcal{A}^+ & 0 \end{pmatrix}
\tag{4.87}
$$

Pursuing the analogy with the definition of H_s in SUSYQM, here too, we can write down a quantity K defined by

$$
\begin{aligned}
K &= \{q, q^+\} \\
&= (q + q^+)^2
\end{aligned}
\tag{4.88}
$$

However K, unlike H_s, is a fourth-order differential operator. The passage from SUSY to SSUSY is thus a transition from $H_s \to K$. One of the purposes in this section is to show that in higher-derivative models there are problems in using Witten index to characterize spontaneous SUSY breaking [66,68].

Let us suppose the existence of an h-operator as a diagonal 2×2 matrix operator that commutes with q and q^+

$$
h = \begin{pmatrix} h_- & 0 \\ 0 & h_+ \end{pmatrix}
\tag{4.89}
$$

$$
[h, q] = 0 = [h, q^+]
\tag{4.90}
$$

It gives rise to the following interwining relations in terms of $\mathcal{A}$ and $\mathcal{A}^+$

$$
\begin{aligned}
h_+ \mathcal{A}^+ &= \mathcal{A}^+ h_- \\
\mathcal{A} h_+ &= h_- \mathcal{A}
\end{aligned}
\tag{4.91}
$$

These are similar in form to (4.85).

We now exploit the above relations to obtain a constraint equation between the functions $p(x)$ and $g(x)$. Indeed if we define h_+ and h_- in terms of the potential v_+ and v_- as

$$
h_\pm = -\frac{1}{2}\frac{d^2}{dx^2} + v_\pm(x)
\tag{4.92}
$$

and substitute (4.86) in (4.91), we obtain

$$
g = p' + 2p^2 - \frac{p''}{4p} + \frac{1}{2}\left(\frac{p'}{2p}\right)^2 + \frac{\mu}{32p^2}
\tag{4.93}
$$

where the dashes denote derivatives with respect to x and μ is an arbitrary real constant. Also $v_\pm(x)$ are given by

$$
v_+ = -2p' + 2p^2 + \frac{p''}{4p} - \frac{1}{2}\left(\frac{p'}{2p}\right)^2 - \frac{\mu}{32p^2} + \lambda
\tag{4.94a}
$$

$$v_- = 2p' + 2p^2 + \frac{p''}{4p} - \frac{1}{2}\left(\frac{p'}{2p}\right)^2 - \frac{\mu}{32p^2} + \lambda \qquad (4.94b)$$

where λ is an arbitrary real constant.

To proceed further let us consider now the possibility of factorizing the operators A and A^+. We write

$$A = \frac{1}{2}bc \qquad (4.95)$$

where b and c are given by the form

$$b = \frac{d}{dx} + U_+(x)$$

$$c = \frac{d}{dx} + U_-(x) \qquad (4.96)$$

From the first equation of (4.86) we can deduce a connection of the functions $U_\pm$ with $p(x)$ and $g(x)$:

$$\begin{aligned} 4p &= U_+ + U_- \\ 2g &= U_+U_- + U'_- \end{aligned} \qquad (4.97)$$

Turning to the quasi-Hamiltonian K we note that for simplicity we can assume it to be a polynomial in h (in the present case, second order in h only). This leads to the picture of the so-called "polynomial SUSY." It must, however, be admitted that a physical interpretation of K is far from clear. Classically, one encounters a fourth-order differential operator while dealing with the problem of an oscillating elastic rod or in the case of a circular plate loaded symmetrically [79,80]. Quantum mechanically the situation is less obvious. However, HSUSSY, unlike the usual SUSQM, allows residual symmetries [64,65]. It may be remarked that coupled channel problems and transport matrix potentials come under the applications of higher derivative schemes [69].

Expressing K as

$$\begin{aligned} K &= h^2 - 2\lambda h + \mu \\ &= (h - \lambda)^2 + \mu - \lambda^2 \end{aligned} \qquad (4.98)$$

we note that

$$[K, h] = 0 \qquad (4.99)$$

A particularly interesting case appears when K is expressed as a perfect square

$$K = (h - \lambda)^2 \tag{4.100}$$

with

$$\lambda = \sqrt{\mu}, \ \mu > 0 \tag{4.101}$$

This means that K can be written as

$$K = \begin{bmatrix} (h_- - \lambda)^2 & 0 \\ 0 & (h_+ - \lambda)^2 \end{bmatrix} \tag{4.102}$$

But K is also given (4.88) which implies

$$\begin{aligned} K &= qq^+ + q^+q \\ &= \begin{pmatrix} \mathcal{A}\mathcal{A}^+ & 0 \\ 0 & \mathcal{A}^+\mathcal{A} \end{pmatrix} \end{aligned} \tag{4.103}$$

As such from (4.102) and (4.103) we are led to

$$\begin{aligned} \mathcal{A}\mathcal{A}^+ &= (h_- - \lambda)^2 \\ \mathcal{A}^+\mathcal{A} &= (h_+ - \lambda)^2 \end{aligned} \tag{4.104}$$

To express the left hand side as a perfect square we see that a constraint of the type

$$cc^+ = b^+b \tag{4.105}$$

can do the desired job. Indeed using (4.105) we find

$$\begin{aligned} (h_- - \lambda)^2 &= \mathcal{A}\mathcal{A}^+ \\ &= \frac{1}{4}bcc^+b^+ \\ &= \frac{1}{4}bb^+bb^+ \\ &= \left(\frac{1}{2}bb^+\right)^2 \end{aligned} \tag{4.106}$$

Similarly,

$$(h_+ - \lambda)^2 = \left(\frac{1}{2}c^+c\right)^2 \tag{4.107}$$

In these words we can factorize $h_\pm$ in a rather simple way

$$\begin{aligned} h_- &= \frac{1}{2}bb^+ + \lambda \\ h_+ &= \frac{1}{2}c^+c + \lambda \end{aligned} \tag{4.108}$$

We now utilize the constraint relation (4.105) by substituting in it the representation (4.96) for b, c and their conjugates. We derive in this way the result

$$U_-^2 + U_-' = U_+^2 - U_+' \qquad (4.109a)$$

When $\mu = 0$ [which implies from (4.101) $\lambda = 0$ as well] the functions $U_\pm$ are explicitly given by

$$U_+ = \frac{p'}{2p} + 2p$$

$$U_- = -\frac{p'}{2p} + 2p \qquad (4.109b)$$

where we have made use of (4.97) and (4.93).

We thus have from (4.108) and (4.105), a triplet of Hamiltonians $(\frac{1}{2}bb^+, \frac{1}{2}cc^+, \frac{1}{2}c^+c)$ of which the middle one playing superpartner to the first and third components. More precisely we find the situation that H is being built up from two standard supersymmetric Hamiltonians namely $\frac{1}{2}(bb^+, cc^+)$ and $\frac{1}{2}(cc^+, c^+c)$.

To inquire into the role of the Witten index in the above scheme we look for the zero-modes of the quasi-Hamiltonian K. These are provided by the equations

$$\mathcal{A}\psi_B = 0, \quad \mathcal{A}^+\psi_F = 0 \qquad (4.110a)$$

It is a straightforward exercise to check that ψ_B and ψ_F are

$$\psi_B = c_B \sqrt{p} \exp\left(-\int_{x_0}^x p\, dt\right)$$

$$\psi_F = c_F \sqrt{p} \exp\left(\int_{x_0}^x p\, dt\right) \qquad (4.110b)$$

where c_B and c_F are constants.

Witten index Δ of (4.8) is thus determined by the asymptotic nature of $p(x)$ that renders ψ_B and ψ_F normalizable. So it is clear that $\Delta \in (-1, 1)$ since the number of vacuum states can be [66] $N_{B,F} = 0, 1$.

The special case corresponding to

$$|x| \to \infty : \ p(x) \to 0, \ \int_{-\infty}^\infty p(x)\, dx < \infty \qquad (4.111)$$

is of interest [68] since here zero modes exist corresponding to both ψ_B and ψ_F. As a result a possible configuration can develop through $N_B = N_B = 1$ implying $\Delta = 0$. The intriguing point is that the vanishing of the Witten index does not imply absence of zero modes and in consequence occurrence of spontaneous breaking of SUSY. On the contrary we have a doubly degenerate zero modes of the operators $\mathcal{A}$ and $\mathcal{A}^+$.

Finally, we address to the more general possibility when the parameter μ is nonzero. For this, let us return to the constraint (4.93). The sign of μ decides whether the algebra is reducible ($\mu < 0$) or not ($\mu > 0$). In the reducible case we find for $\lambda = 0$ and $\nu^2 = -4\mu$ the following features

$$\mu < 0 : h_- = \frac{1}{2}\left(bb^+ + \frac{\nu}{2}\right)$$
$$h_0 = \frac{1}{2}\left(cc^+ + \frac{\nu}{2}\right)$$
$$h_+ = \frac{1}{2}\left(c^+c - \frac{\nu}{2}\right) \tag{4.112}$$

along with

$$U_\pm = \pm\frac{p'}{2p} + 2p \mp \frac{\nu}{8p} \tag{4.113}$$

(4.112) indicates that there exists an intermediate Hamiltonian h_0 which is superpartner to both h_- and h_+. However, if $\mu > 0$ then ν turns imaginary and there can be no hermitean intermediate Hamiltonian.

4.10 Explicit SUSY Breaking and Singular Superpotentials

So far we have considered unbroken and spontaneously broken cases of SUSY. Let us now make a few remarks on the possibility of explicit breaking of SUSY. This is to be distinguished from the spontaneous breaking in that for the explicit breaking the SUSY algebra does not work in the conventional sense in the Hilbert space but rather as an algebra of formal differential operators [81,82]. Explicit breaking of SUSY can be accompained by negative ground state energy with unpairing states in contrast to spontaneously broken SUSY when all levels are paired. Explicit breaking of SUSY can be caused by the presence of singular superpotentials. Note that so far in our discussions we were concerned with continuously differentiable superpotentials which in turn led to nonsingular supercharges and partner supersymmetric Hamiltonians in $(-\infty, \infty)$.

Let us consider now a singular superpotential of the type

$$W(x) = \frac{\nu}{x} - x \tag{4.114}$$

where $\nu \in \mathbb{R}$. With $W(x)$ given above one can easily work out the partner Hamiltonians

$$
\begin{aligned}
H_\pm &= \frac{1}{2}\left(W^2 \mp W'\right) \\
&= \frac{1}{2}\left[-\frac{d^2}{dx^2} + x^2 + \frac{\nu(\nu \mp 1)}{x^2} - (2\nu \mp 1)\right]
\end{aligned}
\tag{4.115}
$$

At the point $\nu = 1$, the component H_+ is found to shed off the singular term and to acquire the form of the oscillator Hamiltonian except for a constant term and to acquire the form of the oscillator Hamiltonian except for a constant term

$$H_+ = \frac{1}{2}\left(-\frac{d^2}{dx^2} + x^2 - 3\right) \tag{4.116}$$

However, H_- is singular.

The interesting point is that if we focus on H_+ we find that it possesses the spectrum

$$E_+^n = \left(n + \frac{1}{2}\right) - \frac{3}{2} \tag{4.117}$$

for $n = 0, 1, 2, \ldots$. For $n = 0$ one is naturally led to a negative ground state energy $E_+^0 = -1$ and in consequence SUSY breaking. Actually SUSY remains broken in the entire interval $\frac{1}{2} < \nu < \frac{3}{2}$ with the ground state energy given by $E_+^0 = -2\nu + 1 < 0$. Note that no negative norm state is associated with $\nu \in (\frac{1}{2}, \frac{3}{2})$. However, $Q|0 >$ turns out to be nonnormalizable and not belonging to the Hilbert space. Indeed it is the nonnormalizability of $Q|0 >$ which causes the ground state to lose its semi-positive definiteness character. Jevicki and Rodrigues [83] have made a detailed analysis of the model proposed by (4.114) and have found several ranges of the coupling $\nu < -\frac{3}{2}$ and $-\frac{3}{2} < \nu < -\frac{1}{2}$ apart from the one we have just now mentioned. In both these cases, however, the ground state energy remains positive.

Casahorran and Nam [81,82] have made further studies on the explicit nature of SUSY breaking. They have obtained a new family of singular superpotentials which include the class of Pöschl-Teller potentials. In particular for the system ($l < 0$)

$$
\begin{aligned}
H_+ &= -\frac{1}{2}\frac{d^2}{dx^2} + \frac{1}{2}\left(1 + |l|^2\right)^2 - \frac{1}{2}|l|\left(|l| - 1\right)\operatorname{sech}^2 x \\
E_+^n &= \frac{1}{2}\left[(1 + |l|)^2 - (|l| - 1 - n)^2\right], \\
&\qquad n = 0, 1, \ldots < |l| - 1
\end{aligned}
\tag{4.118}
$$

and its partner

$$
\begin{aligned}
H_- &= -\frac{1}{2}\frac{d^2}{dx^2} + \frac{1}{2}\left(1 + |l|\right)^2 - \frac{1}{2}|l|\left(|l| + 1\right)\operatorname{sech}^2 x + \operatorname{cosech}^2 x \\
E_-^n &= \frac{1}{2}\left[(1 + |l|)^2 - (|l| - 2 - 2n)^2\right], \\
&\qquad n = 0, 1, \ldots, < \frac{|l|}{2} - 1
\end{aligned}
\tag{4.119}
$$

SUSY can be seen to be explicitly broken for $l < -1$ with unpaired states. Indeed one can see that with $l < -1, H_+$ possesses positive energy eigenstates. The condition for H_- to possess positive energy eigenstates is, however, $l < -2$.

There have also been other proposals in the literature with singular superpotentials. The one suggested by Roy and Roychoudhury, namely [84]

$$
W(x) = -x + \frac{2\nu x}{1 + \nu x^2} + \frac{\mu}{x} + \sum_{i=1}^{n} \frac{2\lambda_i x}{1 + \lambda_1 x^2}
\tag{4.120}
$$

exhibits two negative eigenstates. Occurrence of negative energy states in SUSY models has also been discussed in [85].

4.11   References

[1] F. Cooper and B. Freedman, *Ann. Phys.*, **146**, 262, 1983.

[2] A. Turbiner, *Phys. Lett.*, **B276**, 95, 1992.

[3] E. Witten, *Nucl. Phys.*, **B202**, 253, 1982.

[4] M.A. Shifman, IETP Lectures on Particle Physics and Field Theory, World Scientific, Singapore, 1999.

[5] P.G.O. Freund, *Introduction to Supersymmetry*, Cambridge University Press, Cambridge, 1986.

[6] R. Akhoury and A. Comtet, *Nucl. Phys.*, **B245**, 253, 1984.

[7] A. Das, A Kharev, and V.S. Mathur, *Phys. Lett.*, **B181**, 299, 1986.

[8] C. Xu and Z.M. Zhou, *Int. J. Mod. Phys.*, **A7**, 2515, 1992.

[9] H. Umezawa, *Advanced Field Theory*, **AIP**, 1993.

[10] A.K. Ekert and P.L. Knight, *Am. J. Phys.*, **57**, 692, 1989.

[11] G.J. Milburn, *J. Phys. Math. Genl.*, **A17**, 737, 1984.

[12] B. Bagchi and D. Bhaumik, *Mod. Phys. Lett.*, **A15**, 825, 2000.

[13] S. Cecotti and L. Girardello, *Phys. Lett.*, **110B**, 39, 1982.

[14] S. Cecotti and L. Girardello, *Nucl. Phys.*, **B329**, 573, 1984.

[15] M. Atiyah, R. Bott, and V.K. Patodi, *Inventiones Math.*, **19**, 279, 1973.

[16] C. Callias, *Comm. Math. Phys.*, **62**, 213, 1978.

[17] M. Hirayama, *Prog. Theor. Phys.*, **70**, 1444, 1983.

[18] A.J. Niemi and L.C.R. Wijewardhana, *Phys. Lett.*, **B138**, 389, 1984.

[19] D. Boyanovsky and R. Blankenbecler, *Phys. Rev.*, **D30**, 1821, 1984.

[20] A. Jaffe, A. Lesniewski, and M. Lewenstein, *Ann. Phys.*, **178**, 313, 1987.

[21] D. Bolle, F. Gesztesy, H. Grosse, W. Schweiger, and B. Simon, *J. Math. Phys.*, **28**, 1512, 1987.

[22] A. Kihlberg, P. Salomonson, and B.-S. Skagerstam, *Z. Phys.*, **C28**, 203, 1985.

[23] T. Kato,*Perturbation Theory for Linear Operators*, Springer-Verlag, Berlin, 1984.

[24] K. Fujikawa, *Phys. Rev.*, **A52**, 3299, 1995.

[25] B. Bagchi and P.K. Roy, *Phys. Lett.*, **A200**, 411, 1995.

[26] B. Bagchi and D. Bhaumik, *Oscillator Algebra and Q Deformation*, preprint.

[27] V. Kuryshkin, *Anal. Fnd. Louis de. Broglie*, **5**, 11, 1980.

[28] C. Zachos, in *Deformation Theory and Quantum Groups with Applications to Mathematical Physics*, M. Gerstenhaber and J. Stashef, Eds. AMS, Providence, RI, **134**, 1992.

[29] T. Curtright, D. Fairlie, and C. Zachos, Eds., *Proc Argonne Workshop on Quantum Groups*, World Scientific, Singapore, 1991.

[30] J. Fuchs, *Affine Lie Algebras and Quantum Groups*, Cambridge University Press, Cambridge, 1992.

[31] A. Janussis, G. Brodimas, D. Sourlas, and V. Ziszis, *Lett. Nuovo Cim*, **30**, 123, 1981.

[32] L.C. Biedenharn, *J. Phys. Math. Genl.*, **A22**, L873 1989.

[33] A.J. Macfarlane, *J. Phys. Math. Genl.*, **A22**, 4581, 1989.

[34] C.P. Sun and H.C. Fu, *J. Phys. Math. Genl.*, **A22**, L983, 1989.

[35] V. Drinfeld, *Sev. Math. Dokl.*, **32**, 254, 1985.

[36] M. Jimbo, *Lett. Math. Phys.*, **10**, 63, 1985.

[37] M. Jimbo, *Lett. Math. Phys.*, **11**, 247, 1986.

[38] Y.J. Ng, *J. Phys. Math. Gen.*, **A23**, 1023, 1989.

[39] M. Chaichian, *Phys. Lett.*, **237B**, 401, 1990.

[40] D.I. Fivel, *Phys. Rev. Lett.*, **65**, 3361, 1990.

[41] D.I. Fivel, *J. Phys A: Math. Gel.* **24**, 3575, 1991.

[42] F.J. Narganes - Quijano, *J. Phys A: Math. Genl.*, **24**, 593, 1991.

[43] R.M. Mir-Kasimov, *J. Phys. A: Math. Genl.*, **24**, 4283, 1991.

[44] R. Chakrabarti and R. Jagannathan, *J. Phys. A: Math. Gen.*, **24**, L711, 1991.

[45] J. Beckers and N. Debergh, *J. Phys. A: Math. Gen.*, **24**, L1277, 1991.

[46] E. Papp, *J. Phys. A. Math. Gen.*, **29**, 1795, 1996.

[47] V. Chari and A. Pressley, *A Guide to Quantum Groups*, Cambridge University Press, Cambridge, 1994.

[48] S. Majid, *Foundations of Quantum Group Theory*, Cambridge University Press, Cambridge, 1996.

[49] F. Wilczek and A. Zee, *Phys. Rev. Lett.*, **51**, 2250, 1983.

[50] G.V. Dunne, R. Jackiw, and C.A. Trugenberger, Univ. of Maryland Report CTP 1711, College Park, MD, 1989.

[51] T.L. Curtright and C.K. Zachos, *Phys. Lett.*, **243B**, 237, 1990.

[52] C. Quesne, *Phys. Lett.*, **A153**, 203, 1991.

[53] L. Alvarez - Gaume, C. Gomez, and G. Sierra, *Phys. Lett.*, **220B**, 142, 1989.

[54] J. Wess and B. Zumino, CERN - TH 5697/90, Preprint.

[55] S-R Hao, G-H Li, and J-Y Long, *J. Phys A: Math Gen.*, **27**, 5995, 1994.

[56] V. Spiridonov, *Mod. Phys. Lett.*, **A7**, 1241, 1992.

[57] J. Beckers and N. Debergh, *Phys. Lett.*, **286B**, 290, 1992.

[58] B. Bagchi, *Phys. Lett.*, **309B**, 85, 1993.

[59] B. Bagchi and K. Samanta, *Phys. Lett.*, **A179**, 59, 1993.

[60] K. Fujikawa, L.C. Kwek, and C.H. Oh, *Mod. Phys. Lett.*, **A10**, 2543, 1995.

[61] J.K. Sharma, C.L. Mehta, and E.C.G. Sudarshan, *J. Math. Phys.*, **19**, 2089, 1978.

[62] A.J. Macfarlane, *J. Math. Phys.*, **35**, 1054, 1994.

[63] S. Chaturvedi and V. Srinivasan, *Phys. Rev.*, **A44**, 8024, 1991.

[64] A.A. Andrianov, N.V. Borisov, and M.V. Ioffe, *Phys. Lett.*, **A105**, 19, 1984.

[65] A.A. Andrianov, N.V. Borisov, M.I. Eides, and M.V. Ioffe, *Phys. Lett.*, **A109**, 143, 1985.

[66] A.A. Andrianov, M.V. Ioffe, and V. Spiridonov, *Phys. Lett.*, **A174**, 273, 1993.

[67] A.A. Andrianov, M.V. Ioffe, and D.N. Nishnianidze, *Theor. Math. Phys.*, **A104**, 1129, 1995.

[68] A.A. Andrianov, F. Cannata, J.P. Dedonder, and M.V. Ioffe, *Int. J. Mod. Phys.*, **A10**, 2683, 1995.

[69] A.A. Andrianov, F. Cannata, M.V. Ioffe, and D.N. Nishnianidze, *J. Phys. A: Math Gen.*, 5037, 1997.

[70 A.A. Andrianov, F. Cannata, and M.V. Ioffe, *Mod. Phys. Lett.*, **A11**, 1417, 1996.

[71] B.F. Samsonov, *Mod. Phys. Lett.*, **A11**, 1563, 1996.

[72] V.G. Bagrov and B.F. Samsonov, *Theor. Math. Phys.*, **104**, 1051, 1995.

[73] B. Bagchi, A. Ganguly, D. Bhaumik, and A. Mitra, *Mod. Phys. Lett.*, **A14**, 27, 1999.

[74] B. Bagchi, A. Ganguly, D. Bhaumik, and A. Mitra, *Mod. Phys. Lett.*, **A15**, 309, 2000.

[75] A.A. Andrianov, M.V. Ioffe, and D.N. Nishnianidze, *J. Phys. A: Math. Gen.*, **32**, 4641, 1999.

[76] D.J. Fernandez, *Int. J. Mod. Phys.*, **A12**, 171, 1997.

[77] D.J. Fernandez and V. Hussin, *J. Phys. A. Math. Gen.*, **32**, 3603, (1999).

[78] J.I. Diaz, J. Negro, L.M. Nieto, and O. Rosas - Ortiz, *J. Phys. A. Math. Gen.*, **32**, 8447, 1999.

[79] A.E.H. Love, *A Treatise on the Mathematical Theory of Elasticity*, 4th ed., Dover, NY, 1994.

[80] A. Sommerfeld, *Partial Differential Equations in Physics*, Academic Press, New York, 1949.

[81] J. Casahorran and S. Nam, *Int. J. Mod. Phys.*, **A6**, 2729, 1991.

[82] J. Casahorran and J.G. Esteve, *J. Phys. A. Math. Gen.*, **25**, L347, 1992.

[83] A. Jevicki and J.P. Rodrigues, *Phys. Lett.*, **B146**, 55, 1984.

[84] P. Roy and R. Roychoudhury, *Phys. Lett.*, **A122**, 275, 1987.

[85] P. Roy, R. Roychoudhury, and Y.P. Varshni, *J. Phys. A: Math. Gen.*, **21**, 3673, 1988.

CHAPTER 5

Factorization Method, Shape Invariance

5.1 Preliminary Remarks

As we already know modelling of SUSY in quantum mechanical systems rests in the possibility of factorizing the Schroedinger Hamiltonian. In effect this amounts to solving a nonlinear differential equation for the superpotential that belongs to the Riccati class [see (2.39)]. Not all forms of the Schroedinger equation however meet the solvability criterion, only a handful of potentials exist which may be termed as exactly solvable.

Tracking down solvable potentials is an interesting problem by itself in quantum mechanics [1]. Those which possess normalizable wavefunctions and yield a spectra of energy-levels include the harmonic oscillator, Coulomb, isotropic oscillator, Morse, Pöschl-Teller, Rosen-Morse, and $\mathrm{sec}h^2$ potentials. The forms of these potentials are generally expressible in terms of known functions of algebraic polynomials, exponentials, or trigonometric quanties. Importance of searching for solvable potentials stems from the fact that they very often serve as a springboard for undertaking calculations of more complicated systems. SUSY offers a clue [2,3] to the general nature of solvability in that most of the partner potentials derived from the pair of isospectral Hamiltonians satisfy the condition of shape similarity. In other words the functional forms of the partner potentials are similar except for the presence of the governing parameters in the respective potentials. By imposing the so-called "shape invariance" (SI) or "form invariance" condition [4,5] definite expression for the energy levels can be arrived at in closed forms. Although sufficient, the SI condition is not necessary for the solvability of the Schroedinger equation [6]. However, a number of attempts have been made to look for them by employing the SI condition. Before we take up the SI condition let us review briefly the underlying ideas of the factorization method in quantum mechanics [7-23].

5.2   Factorization Method of Infeld and Hull

The main idea of the factorization method is to replace a given Schroedinger equation, which is a second-order differential equation, by an equivalent pair of first-order equations. This enables us to find the eigenvalues and the normalized eigenfunctions in a far easier manner than solving the original Schroedinger equation directly. Indeed the factorization technique has proven to be a powerful tool in quantum mechanics. The factorization method has a long history dating back to the old papers of Schroedinger [17-19], Weyl [20], Dirac [21], Stevenson [22], and Infeld and Hull (IH) [7,8]. IH showed that, for a wide class of potentials, the factorization method enables one to immediately find the energy spectrum and the associated normalized wave functions.

Consider the following Schroedinger equation

$$-\frac{1}{2}\frac{d^2\psi(x)}{dx^2} + [V(x,c) - E]\,\psi(x) = 0 \qquad (5.1)$$

where we suppose that the potential $V(x,c)$ is given in terms of a set of parameters c. We can think of c as being represented by $c = c_0 + m$, $m = 0,1,2,\ldots$ or by a scaling $c_i = qc_{i-1}, 0 < q < 1, i = 0,1,2,\ldots$ However, any specific form of c will not concern us until later in the chapter.

The factorizability criterion implies that we can replace (5.1) by a set of first-order differential operators A and A^+ such that

$$A(x,c+1)A^+(x,c+1)\psi(x,E,c) = -[E + g(c+1)]\,\psi(x,E,c)$$

$$A^+(x,c)A(x,c)\psi(x,E,c) = -[E + g(c)]\,\psi(x,E,c) \qquad (5.2a,b)$$

To avoid confusion we have displayed explicitly the coordinate x and the parameter c on the wave function ψ and also on the first-order operators A and A^+ which are taken to be

$$A(x,c) = \frac{d}{dx} + W(x,c)$$

$$A^+(x,c) = -\frac{d}{dx} + W(x,c) \tag{5.3}$$

In (5.2) g is some function of c while in (5.3) W is an arbitrary function of x and c.

It is easy to convince oneself that if $\psi(x,E,c)$ is a solution of (5.1) then the two functions defined by $\psi(x,E,c+1) = A^+(x,c+1)$ $\psi(x,E,c)$ and $\psi(x,E,c-1) = A(x,c)\psi(x,E,c)$ are also solutions of the same equation for some fixed value of E. This follows in a straightforward way by left multiplying (5.2a) and (5.2b) by the operators $A^+(x,c+1)$ and $A(x,c)$, respectively. As our notations make the point clear, the solutions have the same coordinate dependence but differ in the presence of the parameters. Moreover the operators A and A^+ are mutually self-adjoint due to $\int_a^b \phi(A^+f)dx = \int_a^b (A\phi)fdx$, f being arbitrary subject to the continuity of the integrands and vanishing of ϕf at the end-points of (a,b).

The necessary and sufficient conditions which the function $W(x,c)$ ought to satisfy for (5.1) to be consistent with the pair (5.2) are

$$\begin{aligned} W^2(x,c+1) + W'(x,c+1) &= V(x,c) - g(c+1) \\ W^2(x,c) - W'(x,c) &= V(x,c) - g(c) \end{aligned} \tag{5.4}$$

Subtraction yields

$$\begin{aligned} W^2(x,c+1) + W'(x,c+1) \\ - \left[W^2(x,c) - W'(x,c) \right] = h(c) \end{aligned} \tag{5.5}$$

where $h(c) = g(c) - g(c+1)$. Eq. (5.5) can also be recast in the form

$$V_-(x,c+1) = V_+(x,c) + \frac{1}{2}h(c) \tag{5.6}$$

where $V_\pm$ can be recognized to be the partner components of the supersymmetric Hamiltonian [see (2.29)]. So the function $W(x)$ in (5.3) essentially plays the role of the superpotential.

IH noted that in order for the factorization method to work the quantity $g(c)$ should be independent of x. Taking as a trial solution

$$W(x, c) = W_0 + cW_1 \tag{5.7}$$

the following constraints emerge from (5.5)

$$
\begin{aligned}
a \neq 0 : W_1^2 + W_1' &= -a^2 \\
W_0' + W_0 W_1 &= -ka, \\
g(c) &= a^2 c^2 + 2kca^2 \\
a = 0 : W_1 &= (x + d)^{-1} \\
W_0' + W_0 W_1 &= b_1 \\
g(c) &= -2bc
\end{aligned}
\tag{5.8}
$$
$$\tag{5.9}$$

where a, b, d and k are constants.

The solution (5.7) alongwith (5.8) lead to various types of factorizations

$a \neq 0$

    Type A:

$$
\begin{aligned}
W_1 &= a \cot a(x + x_0) \\
W_0 &= ka \cot a(x + x_0) + \frac{c}{\sin a(x + x_0)}
\end{aligned}
\tag{5.10}
$$

    Type B:

$$
\begin{aligned}
W_1 &= ia \\
W_0 &= iak + e \exp(-iax)
\end{aligned}
\tag{5.11}
$$

$a = 0$

    Type C:

$$
\begin{aligned}
W_1 &= \frac{1}{x} \\
W_0 &= \frac{bx}{2} + \frac{e}{x}
\end{aligned}
\tag{5.12}
$$

    Type D:

$$
\begin{aligned}
W_1 &= 0 \\
W_0 &= bx + p
\end{aligned}
\tag{5.13}
$$

where x_0, e, and p are contents.

A possible enlargement of the decomposition (5.7) can be made by including an additional term $\frac{W_2}{c}$. This induces two more types of factorizations

Type E:

$$
\begin{aligned}
W_1 &= a \cot a(x + x_0) \\
W_0 &= 0 \\
W_2 &= q
\end{aligned}
\tag{5.14}
$$

Type F:

$$
\begin{aligned}
W_1 &= \frac{1}{x} \\
W_0 &= 0 \\
W_2 &= q
\end{aligned}
\tag{5.15}
$$

where

$$
W = W_0 + cW_1 + \frac{W_2}{c}
\tag{5.16}
$$

and q is a constant.

Each type of factorization determines $W(x, c)$ from the solutions of W_0 and W_1 given above. For Types A-D factorizations, $g(c)$ is obtained from its expression in (5.8) whereas for the cases E and F, $g(c)$ can be determined to be $a^2c^2 - \frac{q^2}{c^2}$ and $-\frac{q^2}{c^2}$, respectively. IH concluded that the above types of factorizations are exhaustive if and only if a finite number of negative powers of c are considered in the expansion of $W(x, c)$.

Concerning the normalizability of eigenfunctions we note that $g(c)$ could be an increasing (class I) or a decreasing (class II) function of the parameter c. So we can set $c = 0, 1, 2, \ldots k$ for each of a discrete set of values $E_k(k = 0, 1, 2, \ldots)$ of E for class I and $c = k, k+1, k+2, \ldots$ for each of a discrete set of values $E_k(k = 0, 1, 2, \ldots)$ of E for class II functions.

Replacing ψ in (5.2) by the form Y_k^c we can express the normalized solutions as

Class I:

$$
Y_k^{c-1} = [g(k+1) - g(c)]^{-\frac{1}{2}} \left[W(x, c) + \frac{d}{dx} \right] Y_k^c
\tag{5.17}
$$

Class II:

$$Y_k^{c+1} = [g(k) - g(c+1)]^{-1/2} \left[W(x, c+1) - \frac{d}{dx} \right] Y_k^c \qquad (5.18)$$

where

$$Y_k^k = A \exp \left(\int W(x, k+1) \right) dx \qquad (5.19)$$

for class I and

$$Y_k^k = B \exp \left(- \int W(x, k) dx \right) \qquad (5.20)$$

for Class II with A and B fixed from $\int_a^b (Y_k^k)^2 dx = 1$.

We do not go into the details of the evaluation of the normalized solutions. Suffice it to note that some of the representative potentials for Types $A - G$ are respectively those of Poschi Teller, Morse, a system of identical oscillators, harmonic oscillator, Rosen-Morse, and generalized Kepler problems. In the next section we shall return to these potentials while addressing the question of SI in SUSYQM.

To summarize, the technique of the factorization method lays down a procedure by which many physical problems can be solved in a unified manner. We now turn to the SI condition which has proved to be a useful concept in tackling the problem of solvability of quantum mechanical systems.

5.3   Shape Invariance Condition

The SI condition was first utilized by Gendenshtein [4] to study the properties of partner potentials in SUSYQM. Taking a cue from the IH result (5.6), we can define SI as follows. If the profiles of $V_+(x)$ and $V_-(x)$ are such that they satisfy the relationship

$$V_-(x, c_0) = V_+(x, c_1) + R(c_1) \qquad (5.21)$$

where the parameter c_1 is some function of c_0, say given by $c_1 = f(c_0)$, the potentials $V_\pm$ are said to be SI. In other words, to be SI the potentials $V_\pm$ while sharing a similar coordinate dependence can at most differ in the presence of some parameters. Note that (5.5) is an equivalent condition to (5.21).

An example will make the definition of SI clear. Let us take

$$W(x) = c_0 \tanh x : W(\infty) = -W(-\infty) = c_0 \qquad (5.22)$$

Then

$$V_{\pm}(x, c_0) = -\frac{1}{2}c_0(c_0 \pm 1)\mathrm{sech}^2 x + \frac{c_0^2}{2} \tag{5.23}$$

But these can also be expressed as

$$V_-(x, c_0) = V_+(x, c_1) + R(c_1), \quad c_1 = c_0 - 1 \tag{5.24}$$

where

$$R(c_1) = \frac{1}{2}\left[c_0^2 - c_1^2\right] \tag{5.25}$$

So the potentials $V_{\pm}$ are SI in accordance with the definition (5.21).

To exploit the SI condition let us assume that (5.21) holds for a sequence of parameters $\{c_k\}, k = 0, 1, 2, \ldots$ where $c_k = f f \ldots k$ times $(c_0) = f^k(c_0)$. Then

$$H_-(x, c_k) = H_+(x, c_{k+1}) + R(c_k) \tag{5.26}$$

where $k = 0, 1, 2, \ldots$ and we call $H^{(0)} = H_+(x, c_0), H^{(1)} = H_-(x, c_0)$. Writing $H^{(m)}$ as

$$\begin{aligned}
H^{(m)} &= -\frac{1}{2}\frac{d^2}{dx^2} + V_+(x, c_m) + \sum_{k=1}^{m} R(c_k) \\
&= H_+(x, c_m) + \sum_{k=1}^{m} R(c_k)
\end{aligned} \tag{5.27}$$

it follows on using (5.26) that

$$H^{(m+1)} = H_-(x, c_m) + \sum_{k=1}^{m} R(c_k) \tag{5.28}$$

Thus we are able to set up a hierarchy of Hamiltonians $H^{(m)}$ for various m values. Now according to the principles of SUSYQM highlighted in Chapter 2, H_+ contains the lowest state with a zero-energy eigenvalue. It then transpires from (5.27) that the lowest energy level of $H^{(m)}$ has the value

$$E_0^{(m)} = \sum_{k=1}^{m} R(c_k) \tag{5.29}$$

It is also not too difficult to realize [5] that because of the chain $H^{(m)} \to H^{(m-1)} \ldots \to H^{(1)}(\equiv H_-) \to H^{(0)}(\equiv H_+)$, the nth member in this sequence carries the nth level of the energy spectra of $H^{(0)}$ (or H_+), namely

$$E_n^{(+)} = \sum_{k=1}^{n} R(c_k), \quad E_0^{(+)} = 0 \qquad (5.30)$$

Moreover if $\psi_0^{(+)}(x, c_m)$ is to represent the ground-state wave function for $H^{(m)}$ then the nth wave function for $H_+(x, c_0)$ can be constructed from it by repeated applications of the operator A^+. To establish this we note from (2.59) that $\psi_{n+1}^+ = (2E_n^-)^{-\frac{1}{2}} A^+ \psi_n^-$ and that for SI potentials $\psi_n^{(-)}(x, c_0) = \psi_n^+(x, c_1)$. So we can write

$$\psi_{n+1}^+(x, c_0) = (2E_n^-)^{-1/2} A^+(x, c_0) \psi_n^+(x, c_1) \qquad (5.31)$$

In the presence of n parameters $c_0, c_1, \ldots c_n$, repeated use of (5.31) gives the result

$$\psi_n^+(x, c_0) = N A^+(x, c_0) A^+(x, c_1) \ldots A^+(x, c_{n-1}) \psi_0^+(x, c_n) \qquad (5.32)$$

where N is a constant. These correspond to the energy eigenfunctions of $H_+(x, c_0)$.

Let us now return to the example (5.22). We rewrite (5.24) as

$$V_-(x, c_0) = V_+(x, c_0 - 1) + \frac{1}{2}c_0^2 - \frac{1}{2}(c_0 - 1)^2 \qquad (5.33)$$

and note that we can generate c_k from c_0 as $c_k = c_0 - k$. Therefore the levels of $V_+(x, c_0)$ are given by

$$\begin{aligned}
E_n^+ &= \sum_{k=1}^{n} R(c_k) \\
&= \frac{1}{2} \sum_{k=1}^{n} \left(c_0^2 - c_k^2 \right) \\
&= \frac{1}{2} \left(c_0^2 - c_n^2 \right) \\
&= \frac{1}{2} \left[c_0^2 - (c_0 - n)^2 \right] \qquad (5.34)
\end{aligned}$$

On the other hand, the ground state wave function for $H_+(x, c_0)$ may be obtained from (2.56b) using the form of $W(x)$ in (5.22). It turns out to be proportional to $\mathrm{sech}x$.

E_n^+ for V_+ being obtianed from (5.34) we can easily calculate the energy levels E_n of the potential

$$V(x) = -\beta \operatorname{sech}^2 x \qquad (5.35)$$

with $\beta = \frac{1}{2}c_0(c_0 + 1)$ derived from (5.23). We find

$$E_n = E_n^+ - \frac{1}{2}c_0^2 = -\frac{1}{2}(c_0 - n)^2 \qquad (5.36)$$

where c_0 can be expressed in terms of the coefficient β of $V(x)$.

In Table 5.1 we furnish a list of solvable potentials which are SI in the sense of (5.26). It is worth noting that the well-known potentials such as the Coulomb, the oscillator, Poschl-Teller, Eckart, Rosen-Morse, and Morse, all satisfy the SI condition. The forms of these potentials are also consistent with the following ansatz [6] for the superpotential $W(x, c)$

$$W(x, c) = (a + b)p(x) + \frac{q(x)}{a + b} + r(x) \qquad (5.37)$$

where $c = f(a)$ and b is a constant. Substituting (5.37) into (5.21) it follows that the case $p(x) = q(x) = 0$ leads to the one-dimensional harmonic oscillator, the case $q(x) = 0$ leads to the three-dimensional oscillator and the Morse while the case $r(x) = 0$ along with $q(x) = $ constant leads to the Rosen-Morse, the Coulomb, and the Eckart potentials. Note that the Rosen-Morse potential includes as a particular case $(B = 0)$ the Poschl-Teller potential.

The SI condition has yielded new potentials for a scaling ansatz of the change of parameters as well [24]. With $c_1 = f(c_0)$, let us express (5.21) in terms of the superpotential $W(x)$. We have the form

$$W^2(x, c_0) + W'(x, c_0) = W^2(x, c_1) - W'(x, c_1) + R(c_0) \qquad (5.38)$$

The scaling ansatz deals with the proposition

$$c_1 = qc_0 \qquad (5.39)$$

Table 5.1

A list of **SI** potentials with the parameters explicitly displayed. In the presence of m and $\hbar$, $V_+(x)$ is defined as $V_+(x) = \frac{1}{2}\left(W^2 - \frac{\hbar}{\sqrt{m}}W'\right)$. The variables x and r run between $-\infty < x < \infty$ and $0 < r < \infty$. The results in this table are consistent with the list provided in Ref. [4]. The ground state wave function can be calculated using $\psi_0(x) = \exp\left(-\frac{\sqrt{m}}{\hbar}\int^x W(y)dy\right)$.

Potential	$V_+(x)$	Shape-invariant Parameters		
		c_0	c_1	$R(c_1)$
Shifted Oscillator	$\frac{1}{2}m\omega^2\left(x - \sqrt{\frac{2}{m}}\frac{a}{\omega}\right)^2$	ω	ω	$\hbar\omega$
Isotropic Oscillator (3 dim.)	$\frac{1}{2}m\omega^2 r^2 + \frac{l(l+1)\hbar^2}{2mr^2}$ $-\left(l+\frac{3}{2}\right)\hbar\omega$	l	$l+1$	$2\hbar\omega$
Coulomb	$-\frac{e^2}{r} + \frac{l(l+1)\hbar^2}{2mr^2} + \frac{me^4}{2(l+1)^2\hbar^2}$	l	$l+1$	$\frac{me^4}{2\hbar^2}\left\{\begin{array}{c}\frac{1}{(c_0+1)^2}\\-\frac{1}{(c_1+1)^2}\end{array}\right\}$
Rosen-Morse I	$A^2 + \frac{B^2}{A^2} + 2B\tanh\alpha x$ $-A\left(A + \frac{\alpha\hbar}{\sqrt{2m}}\right)\text{sech}^2\alpha x$	A	$A - \frac{\alpha\hbar}{\sqrt{2m}}$	$c_0^2 - c_1^2$ $+B^2\left(\frac{1}{c_0^2} - \frac{1}{c_1^2}\right)$
Rosen-Morse II	$A^2 + \left(B^2 + A^2 + \frac{A\alpha\hbar}{\sqrt{2m}}\right)\times$ $\text{cosech}^2\alpha r$ $-B\left(2A + \frac{\alpha\hbar}{\sqrt{2m}}\right)\times$ $\coth\alpha r\,\text{cosech}\alpha r$	A	$A - \frac{\alpha\hbar}{\sqrt{2m}}$	$c_0^2 - c_1^2$

Table 5.1 (continued)

Potential	$W(x)$	Energy Levels	$\psi_0(x)$
Shifted Oscillator	$\sqrt{m}\,\omega x - \sqrt{2}\,a$	$n\hbar\omega$	$\exp\left[-\frac{m\omega}{2\hbar}\left(x - \sqrt{\frac{2}{m}}\,\frac{a}{\omega}\right)^2\right]$
Isotropic Oscillator (3 dim.)	$\sqrt{m}\,\omega r - \frac{(l+1)\hbar}{\sqrt{m}\,r^2}$	$2n\hbar\omega$	$r^{l+1}\exp\left(-\frac{m\omega r^2}{2\hbar}\right)$
Coulomb	$\sqrt{m}\,\frac{e^2}{(l+1)\hbar} - \frac{(l+1)\hbar}{\sqrt{m}\,r^2}$	$\frac{me^4}{2\hbar^2}\left[\frac{1}{(l+1)^2} - \frac{1}{(n+l+1)^2}\right]$	$r^{l+1}\exp\left(-\frac{me^2 r}{(l+1)\hbar}\right)$
Rosen-Morse I	$\sqrt{2}\left(A\tanh\alpha x + \frac{B}{A}\right)$	$(\text{sech}\,\alpha x)^{\frac{\sqrt{2mA}}{\alpha\hbar}}\times$ $+B^2\left\{\frac{1}{A^2} - \left(A - \frac{n\hbar\alpha}{\sqrt{2m}}\right)^2\right\}$	$\exp\left(-\frac{\sqrt{2m}\,Bx}{A\hbar}\right)$
Rosen-Morse II	$\sqrt{2}\left(A\coth\alpha r - B\,\text{cosech}\,\alpha r\right)$ $A < B$	$A^2 - \left(A - \frac{n\hbar\alpha}{\sqrt{2m}}\right)^2$	$(\sin har)^{\frac{\sqrt{2m}}{\hbar\alpha}} + (B-A)$ $(1+\cos har)^{\frac{\sqrt{2m}}{\hbar\alpha}\,n}$

Table 5.1 (continued)

Potential	$V_+(x)$	Shape-invariant Parameters		
		c_0	c_1	$R(c_1)$
Eckart-I	$A^2 + \frac{B^2}{A^2} - 2B\coth\alpha r$ $+ A\left(A - \frac{\alpha\hbar}{\sqrt{2m}}\right)\operatorname{cosech}^2\alpha r$	A	$A + \frac{\alpha\hbar}{\sqrt{2m}}$	$c_0^2 - c_1^2$ $+B^2\left(\frac{1}{c_0^2} - \frac{1}{c_1^2}\right)$
Eckart -II	$-A^2 + \left[B^2 + A\left(A - \frac{\hbar\alpha}{\sqrt{2m}}\right)\right]\operatorname{cosec}^2\alpha x$ $- B\left(2A - \frac{\hbar\alpha}{\sqrt{2m}}\right)\operatorname{cosec}\alpha x \cot\alpha x$ $(0 \leq \alpha x \leq \pi, A > B)$	A	$A + \frac{\alpha\hbar}{\sqrt{2m}}$	$c_1^2 - c_0^2$
Poschl-Teller-I	$-(A+B)^2 + A\left(A - \frac{\hbar\alpha}{\sqrt{2m}}\right)\sec^2\alpha x$ $+ B\left(B - \frac{\hbar\alpha}{\sqrt{2m}}\right)\operatorname{cosec}^2\alpha x$	(A, B)	$\left(A + \frac{\alpha\hbar}{\sqrt{2m}},\right.$ $\left.B + \frac{\alpha\hbar}{\sqrt{2m}}\right)$	$\left(A + B + \frac{2\alpha\hbar}{\sqrt{2m}}\right)^2$ $-(A+B)^2$
Poschl-Teller-II	$(A - B)^2 - A\left(A + \frac{\hbar\alpha}{\sqrt{2m}}\right)\operatorname{sech}^2\alpha r$ $+ B\left(B - \frac{\alpha\hbar}{\sqrt{2m}}\right)\operatorname{cosech}^2\alpha r$	(A, B)	$\left(A - \frac{\alpha\hbar}{\sqrt{2m}},\right.$ $\left.B + \frac{\alpha\hbar}{\sqrt{2m}}\right)$	$(A - B)^2 - \left(A\right.$ $\left. -B - \sqrt{\frac{2}{m}}\alpha\hbar\right)^2$
Morse - I	$A^2 + B^2 e^{-2\alpha x} - 2Be^{-\alpha x} \times$ $\left(A + \frac{\alpha\hbar}{2\sqrt{2m}}\right)$	A	$A - \frac{\hbar\alpha}{\sqrt{2m}}$	$c_0^2 - c_1^2$
Hyperbolic	$A^2 + \left[B^2 - A\left(A + \frac{\alpha\hbar}{\sqrt{2m}}\right)\right]\operatorname{sech}^2\alpha x$ $+ B\left(2A + \frac{\alpha\hbar}{\sqrt{2m}}\right)\operatorname{sech}\alpha x \tanh\alpha x$	A	$A - \frac{\alpha\hbar}{\sqrt{2m}}$	$c_0^2 - c_1^2$
Trigonometric	$-A^2 + \frac{B^2}{A^2} + A\left(A + \frac{\hbar\alpha}{\sqrt{2m}}\right)\operatorname{cosec}^2\alpha x$ $- 2B\cot\alpha x$	A	$A - \frac{\hbar\alpha}{\sqrt{2m}}$	$c_1^2 - c_0^2$ $+B^2\left(\frac{1}{c_0^2} - \frac{1}{c_1^2}\right)$

Table 5.1 (continued)

Potential	$W(x)$	Energy Levels	$\psi_0(x)$
Eckart-I	$-\sqrt{2}\left(A\coth\alpha r - \frac{B}{A}\right)$ $(B > A^2)$	$A^2 - \left(A + \frac{n\hbar\alpha}{\sqrt{2m}}\right)^2$ $+ B^2\left[\frac{1}{A^2} - \frac{1}{\left(A + \frac{n\alpha\hbar}{\sqrt{2m}}\right)^2}\right]$	$(\sinh\alpha r)^{\frac{\sqrt{2m}A}{\alpha\hbar}}$ $\exp\left(-\frac{\sqrt{2m}Br}{A\hbar}\right)$
Eckart -II	$\sqrt{2}(-A\cot\alpha x + B\csc\alpha x)$ $(0 \le \alpha x \le \pi, A > B)$	$\left(A + \frac{n\hbar\alpha}{\sqrt{2m}}\right)^2 - A^2$	$(\sin\alpha x)^{\frac{\sqrt{2m}}{\hbar\alpha}(A+B)}$ $(1-\cos\alpha x)^{\frac{\sqrt{2m}}{\hbar\alpha}B}$
Poschl–Teller-I	$\sqrt{2}(A\tan\alpha x - B\cot\alpha x)$ $\left(0 \le \alpha x \le \frac{\pi}{2}\right)$	$\left(A + B + \frac{2\alpha\hbar}{\sqrt{2m}}\right)^2$ $-(A+B)^2$	$(\sin\alpha x)^{\frac{\sqrt{2m}}{\alpha\hbar}B} \times$ $(\cos\alpha x)^{\frac{\sqrt{2m}}{\alpha\hbar}A}$
Poschl–Teller-II	$\sqrt{2}(A\tanh\alpha r - B\coth\alpha r)$ $(B < A)$	$(A-B)^2$ $-\left(A - B - \sqrt{\frac{2}{m}}\,n\alpha\hbar\right)^2$	$(\sinh\alpha r)^{\frac{\sqrt{2m}}{\alpha\hbar}B}$ $(\cosh\alpha r)^{\frac{\sqrt{2m}}{\alpha\hbar}A}$
Morse - I	$\sqrt{2}(A - Be^{-\alpha x})$	$A^2 - \left(A - \frac{n\alpha\hbar}{\sqrt{2m}}\right)^2$	$\exp\left\{-\frac{\sqrt{2m}}{\hbar}\right.$ $\left.\left(Ax + \frac{B}{\alpha}e^{-\alpha x}\right)\right\}$
Hyperbolic	$\sqrt{2}(\tanh\alpha x + B\sec h\alpha x)$	$A^2 - \left(A - \frac{n\alpha\hbar}{\sqrt{2m}}\right)^2$	$(\sec h\alpha x)^{\frac{\sqrt{2m}}{\alpha\hbar}A} \times$ $\exp\left\{-\frac{2\sqrt{2m}B}{\alpha\hbar}\right.$ $\left.\tan^{-1}(e^{\alpha x})\right\}$
Trigonometric	$\sqrt{2}\left(A\cot\alpha x - \frac{B}{A}\right)$ $-2B\cot\alpha x$	$-A^2 + \left(A - \frac{n\hbar\alpha}{\sqrt{2m}}\right)^2$ $+\frac{B^2}{A^2} - \frac{B^2}{\left(A - \frac{n\hbar\alpha}{\sqrt{2m}}\right)^2}$	$(\csc\alpha x)^{\frac{\sqrt{2m}}{\alpha\hbar}A} \times$ $\exp\left\{\frac{\sqrt{2m}B}{\hbar A}x\right\}$

with $q \in (0, 1)$, a fractional quantity. The parameter q in effect yields a deformation of quantum mechanics affected by the q-parameter. As already alluded to in Chapter 4 such a deformation is called q-deformation.

We now consider expansions of $W(x)$ and $R(c_0)$ in a manner

$$W(x, c_0) = \sum_{k=0}^{\infty} t_k(x) c_0^k \qquad (5.40)$$

$$R(c_0) = \sum_{k=0}^{\infty} R_k c_0^k \qquad (5.41)$$

Substituting (5.41) and (5.40) into (5.38) and matching powers of c_0 gives a first-order differential equation for $t_k(x)$

$$\frac{dt_n(x)}{dx} + 2\left[\xi_n t_0(x)\right] t_n(x) = \xi_n \left[\rho_n - \sum_{k=1}^{n-1} t_k(x) t_{n-k}(x)\right] \qquad (5.42)$$

where

$$R_n \equiv (1 - q^n)\rho_n,$$

$$t_0(x) = \frac{1}{2} R_0 x + \lambda,$$

$$\xi_n \equiv (1 - q^n)/(1 + q^n) \qquad (5.43)$$

with $n = 1, 2, \ldots$ and λ is a constant.

The solution of (5.42) corresponding to $t_0 = 0$ is

$$t_n(x) = \xi_n \int \left[\rho_n - \sum_{k=1}^{n-1} t_k(x) t_{n-k}(x)\right] dx \qquad (5.44)$$

Notice that for $t_0 = 0$, both R_0 and λ are vanishing.

To see how the scaling ansatz works consider a nontrivial situation when $\rho_n = 0$ for $n \geq 3$. From the solution (5.44) we can easily derive

$$t_1(x) = \xi_1 \rho_1 x$$

$$t_2(x) = \xi_2 \rho_2 x - \frac{1}{3}\xi_1^2 \rho_1^2 \xi_2 x^3$$

$$t_3(x) = -\frac{2}{3}\xi_1 \rho_1 \xi_2 \rho_2 \xi_3 x^3$$

$$ + \frac{2}{15}\xi_1^3 \rho_1^3 \xi_2 \xi_3 x^5 \qquad (5.45)$$

These indicate $W(x)$ to be an odd function in x (unbroken SUSY) so tht $V_+(x)$ is symmetric.

Using now (5.30) and (5.41) we find

$$E_n^+(c_0) = R_1 c_0 \frac{1-q^n}{1-q}$$

$$+ R_2 c_0^2 \frac{1-q^{2n}}{1-q^2} \qquad (5.46)$$

which may be interpreted to correspond to a deformed spectra. The ground state wave function turns out as

$$\psi_0^+(x, c_0) = \exp\left[-ax^2 + bx^4 + O(x^6)\right] \qquad (5.47)$$

where

$$a = -\frac{1}{2}\left(\xi_1\rho_1 c_0 + \xi_2\rho_2 c_0^2\right),$$

$$b = \frac{1}{12}\left[\xi_2(\xi_1\rho_1 c_0)^2 + 2\xi_3(\xi_1\rho_1 c_0)(\xi_2 r_2 c_0^2)\right.$$

$$\left. +\xi_4(\xi_2\rho_2 c_0^2)^2\right] \qquad (5.48)$$

A different set of ansatz for $W(x, c)$ was proposed by Shabat and Yamilov [25] in terms of an index $k, k \in \mathbb{Z}$, by treating (5.38) as an infinite-dimensional chain and truncating it at a suitable point in an endeavor to look for related potentials.

Translating (5.38) into a chain of coupled Riccati equations involving the index k we have

$$W_k^2(x) + W_k'(x) - W_{k+1}^2(x) + W_{k+1}'(x) = R_k \qquad (5.49)$$

We may impose upon W_k and R_k the following cyclic property

$$W_k(x) = W_{k+N}(x)$$
$$R_k = R_{k+N} \qquad (5.50)$$

where N is a positive integer.

The case $N = 2$ may be worked out easily which is guided by the following equations

$$W_1'(x) + W_2'(x) + W_1^2(x) - W_2^2(x) = R_1$$
$$W_2'(x) + W_1'(x) + W_2^2(x) - W_1^2(x) = R_2 \qquad (5.51)$$

The above equations may be reduced to the forms

$$
\begin{aligned}
2W_1^2(x) - 2W_2^2(x) &= R_1 - R_2 \\
2W_1'(x) - 2W_2'(x) &= R_1 + R_2
\end{aligned}
\tag{5.52}
$$

which when solved give

$$
\begin{aligned}
W_1(x) &= \frac{\delta_1}{x} + \mu_1 \\
W_2(x) &= \frac{\delta_2}{x} + \mu_2
\end{aligned}
\tag{5.53}
$$

where

$$
\delta_1 = -\delta_2 = \frac{1}{2}\left(\frac{R_1 + R_2}{R_1 - R_2}\right)
\tag{5.54}
$$

$$
\mu_1 = \mu_2 = \frac{1}{\mu}(R_1 + R_2)
\tag{5.55}
$$

So the $N = 2$ case gives us the model of conformal quantum mechanics [26].

The case $N = 3$ is represented by the equation

$$
\begin{aligned}
W_1'(x) + W_2'(x) + W_1^2(x) - W_2^2(x) &= R_1 \\
W_2'(x) + W_3'(x) + W_2^2(x) - W_3^2(x) &= R_2 \\
W_3'(x) + W_1'(x) + W_3^2(x) - W_1^2(x) &= R_3
\end{aligned}
\tag{5.56}
$$

A set of solutions for $W_i(x)$, $i = 1, 2, 3$ satisfying (5.56) has the form

$$
\begin{aligned}
W_1(x) &= \frac{1}{2}\omega x + f(x) \\
W_{2,3}(x) &= -\frac{1}{2}f(x) \mp \frac{1}{2f(x)}\left[f'(x) + R_2\right]
\end{aligned}
\tag{5.57}
$$

where the function $f(x)$ needs to satisfy a nonlinear differential equation

$$
\begin{aligned}
2f'' &= \frac{f'^2}{f} + 3f^3 + 4\omega x f^2 \\
&\quad + \left[\omega^2 x^2 + 2(R_3 - R_1)\right]f - \frac{k_2^2}{f}
\end{aligned}
\tag{5.58}
$$

In this way the $N = 3$ case constrains $W(x)$ to depend on the solution of the Painlevé-IV equation yielding transcendental potentials [27]. We thus have a nice interplay between the SI condition on the one hand and Painlevé transcendent on the other.

5.4   Self-similar Potentials

Self-similar potentials have also been investigated within the framework of (5.28). Shabat [28] considered the following self-similarity constraint on the superpotential $W(x)$ guided by the index j

$$W(x, j) = q^j W(q^j x) \tag{5.59}$$
$$E_j = q^{2j} k, \ k > 0 \tag{5.60}$$

and $q \in (0, 1)$. In terms of j the SI condition (5.38) reads

$$W^2(x, j) + W'(x, j) - W^2(x, j+1) + W'(x, j+1) = R(j) \tag{5.61}$$

Now (5.59) is a solution of (5.61) if

$$W^2(x) + W'(x) - q^2 W^2(qx) + q W'(qx) = R \tag{5.62}$$

(5.62) is the condition of self-similarity [29-31]. One can verify that (5.62) can be justified by a q-deformed Heisenberg-Weyl algebra

$$\mathcal{A}\mathcal{A}^+ - q^2 \mathcal{A}^+ \mathcal{A} = R \tag{5.63}$$

where $\mathcal{A}$ and $\mathcal{A}^+$ are defined by

$$\mathcal{A} = T_q^{-1} \left(\frac{d}{dx} + W \right)$$
$$\mathcal{A}^+ = \left(-\frac{d}{dx} + W \right) T_q \tag{5.64}$$

and T_q operates according to

$$T_q f(x) = \sqrt{q} f(qx), \ T_q^{-1} = T_{q^{-1}} \tag{5.65}$$

Such deformed operators as $\mathcal{A}$ and $\mathcal{A}^+$ in (5.64) give rise to a q-deformed SUSYQM.

Solutions to (5.62) can be sought for by employing a power series for $W(x)$ and looking for symmetric potentials

$$W(x) = \sum_{j=1}^{\infty} a_j x^{2j-1} \tag{5.66}$$

Substitution of (5.66) in (5.62) results in the series

$$a_j = \frac{1 - q^{2j}}{1 + q^{2j}} \frac{1}{2j - 1} \sum_{k=1}^{j-1} a_{j-k} a_k \qquad (5.67)$$

with

$$a_1 = R/(1 + q^2) \qquad (5.68)$$

Thus as $q \to 0$ we get Rosen-Morse, as $R \propto q \to \infty$ we get Pöschl-Teller, as $q \to 1$ we get harmonic oscillator, and as $q \to 0$ and $R = 0$ we get the radial potential. Finally, we may point out that the $j = 1$ case of (5.67) is in conformity with $n = 1$ solution of (5.44) with the replacement $q \to q^2$ in it and setting $\rho_n = 0$ for $n \geq 2$.

To conclude, it is interesting to note that, using (5.3), we can put (5.61) in the form

$$A(j)A^+(j) = A^+(j + 1)A(j + 1) + R(j) \qquad (5.69)$$

This facilitates dealing with q-deformed coherent states [32,33] associated with the self-similar potentials.

5.5   A Note On the Generalized Quantum Condition

In this section we consider the possibility of replacing the usual quantum condition (2.6) by a more general one [34-36]

$$[\tilde{b}, \tilde{b}^+] = 1 + 2\nu K \qquad (5.70)$$

where $\nu \epsilon \mathbb{R}$ and K are indempotent operators $K^2 = 1$ such that $\{\tilde{b}, K\} = \{\tilde{b}^+, K\} = 0$. The above condition results from the fact that the Heisenberg equations of motion for the one-dimensional oscillator admit an extended class of commutation relations. More recently, (5.70) has been found relevant [37-39] in the context of integrable models. The above generalized condition is also consistent with the Calogero model [40].

A plausible set of representations for $\tilde{b}$ and $\tilde{b}^+$ obeying (5.70) may be worked out to be

$$\tilde{b} = \frac{1}{\sqrt{2}} \left[x + ip - \frac{\nu K}{x} \right]$$

$$\widetilde{b}^+ = \frac{1}{\sqrt{2}}\left[x - ip + \frac{\nu K}{x}\right] \tag{5.71}$$

(5.71) goes over to (2.2) when ν is set equal to zero.

For solvable systems admitting of SUSY, one can modify (5.70) even further in terms of a superpotential W [41]:

$$[\widetilde{b}, \widetilde{b}^+] = \frac{dW}{dx} + 2\nu K \tag{5.72}$$

For an explicit realization of (5.72), K can be chosen to be σ_3.

The representations for $\widetilde{b}$ and $\widetilde{b}^+$ consistent with (5.72) are

$$\widetilde{b} = \frac{1}{\sqrt{2}}\left(W + \frac{d}{dx}\right)\sigma_1 + \frac{i\nu}{W\sqrt{2}}\sigma_2$$

$$\widetilde{b}^+ = \frac{1}{\sqrt{2}}\left(W - \frac{d}{dx}\right)\sigma_1 - \frac{i\nu}{W\sqrt{2}}\sigma_3 \tag{5.73}$$

where $W(x)$ is restricted to an odd function of x ensuring $\exp(-\int^x W(y)dy) \to 0$ as $x \to \pm\infty$.

Using the expression for the supersymmetric Hamiltonian in the form $2H_s = \left\{\widetilde{b}, \widetilde{b}^+\right\} + \left[\widetilde{b}, \widetilde{b}^+\right]K$, the partner Hamiltonians may be deduced as

$$\begin{aligned} H_\pm &= \frac{1}{2}\left[-\frac{d^2}{dx^2} + W^2 \mp W'\right] + \frac{\nu}{2}\left[\frac{\nu}{W^2} - 2 \mp \frac{W'}{W^2}\right] \\ &= \frac{1}{2}\left(-\frac{d^2}{dx^2} + \omega^2 \pm \omega'\right) \end{aligned} \tag{5.74}$$

where $\omega = \frac{\nu}{W} - W$. So we see that ω is always singular except when $\nu = 0$.

As an application of the scheme (5.74) let us consider the harmonic oscillator case $W = x$. This implies $\omega = \frac{\nu}{x} - x$ which may be recognized to be a SI singular potential, ν playing the role of a coupling constant. The partner Hamiltonians induced are

$$H_\pm = -\frac{1}{2}\frac{d^2}{dx^2} + \frac{1}{2}x^2 + \frac{\nu(\nu \mp 1)}{2x^2} - \frac{1}{2}(2\nu \mp 1) \tag{5.75}$$

It has been remarked in the previous chapter that in the interval $\frac{1}{2} < \nu < \frac{3}{2}$ the unpairing of states is accompanied by [42] a unique energy state that may be negative. We wish to remark that generalized conditions such as (5.70) or (5.72) inevitably give rise to singular Schroedinger potentials.

Finally, we may note that the operator K can also be represented by the Klein operator [43] or the parity operator [44-46]. But these forms are not conducive to the construction of supersymmetric models.

5.6   Nonuniqueness of the Factorizability

In the previous sections we have shown how the factorization method along with the SI condition help us to determine the energy spectra and the wave functions of exactly solvable potentials. However, one particular feature worth examining is the nonuniqueness [47-49] of the factorizability of a quantum mechanical Hamiltonian. We illustrate this aspect by considering the example of the harmonic oscillator whose Hamiltonian reads

$$H = -\frac{1}{2}\frac{d^2}{dx^2} + \frac{1}{2}x^2 \tag{5.76}$$

where we have set $\omega = 1$. H can be written as

$$H = b^+ b + \frac{1}{2} \tag{5.77}$$

where $[b, b^+] = 1$, $b = \left(\frac{d}{dx} + x\right)/\sqrt{2}$ and $b^+ = \left(-\frac{d}{dx} + x\right)/\sqrt{2}$. Further $Hb^+ = b^+(H+1)$, $Hb = b(H-1)$ and the ground-state wave functions ψ_0 can be extracted from $b\psi_0 = 0$ leading to $\psi_0 = c_0 e^{-x^2/2}$ (c_0 is a constant). Further, higher-level wave functions are obtained using $\psi_n = c_n(b^+)^n \psi_0$ (c_n are constants), $n = 1, 2, \ldots$

However, the representation of the factors denoted by b and b^+ are by no means unique. Indeed we can also express (5.77) as

$$H + \frac{1}{2} = \left[2^{-1/2}\left(\frac{d}{dx} + \alpha(x)\right)2^{-1/2}\left(-\frac{d}{dx} + \alpha(x)\right)\right]$$
$$= (b')(b')^+ \tag{5.78}$$

where

$$b' = \left[\frac{d}{dx} + \alpha(x)\right]/\sqrt{2}$$

$$(b')^+ = \left[-\frac{d}{dx} + \alpha(x) \right] / \sqrt{2} \tag{5.79}$$

and we have set $\alpha(x) = x + \beta(x)$, $\beta \neq 0$. A simple calculation gives

$$\beta(x) = \psi_0^{-2} \left[K + \int^x \psi_0^{-2}(y) dy \right]^{-1} \tag{5.80}$$

where K is a constant. Although $\beta(x)$ is not expressible in a closed form for which ψ_0 is required to be an inverse-square integrable function, it is clear from (5.79) that we can define a new Hamiltonian H' given by

$$H' = (b')^+ (b') + \frac{1}{2} \tag{5.81}$$

which has a spectra coinciding with that of the harmonic oscillator (see below) but under the influence of a different potential

$$V'(x) = \frac{x^2}{2} - \frac{d}{dx} \left[e^{-x^2} \left(K + \int_0^x e^{-y^2} dy \right)^{-1} \right] \tag{5.82}$$

$V'(x)$ is singularity-free for $|k| > \sqrt{\pi}/2$ and behaves like $V(x)$ asymptotically.

To establish that the spectra of H and H' coincide we note that

$$
\begin{aligned}
H'(b')^+ &= \left[(b')^+ b' + \frac{1}{2} \right] (b')^+ \\
&= (b')^+ \left[b'(b')^+ + \frac{1}{2} \right] \\
&= (b')^+ (H + 1) \tag{5.83}
\end{aligned}
$$

from (5.78). Further

$$
\begin{aligned}
H' \phi_n &= H'(b')^+ \psi_{n-1} \\
&= (b')^+ (H + 1) \psi_{n-1} \\
&= (b')^+ \left(n + \frac{1}{2} \right) \psi_{n-1} \\
&= \left(n + \frac{1}{2} \right) \phi_n \tag{5.84}
\end{aligned}
$$

where $\phi_n = (b')^+ \psi_{n-1}$ and ψ_n are those of (5.76), $n = 1, 2, \ldots$ Hence we conclude that both H and H' share a similar energy spectra.

So nonuniqueness of factorization allows us to construct a new class of potentials different from the harmonic oscillator but possesses the oscillator spectrum. The nonuniqueness feature of factorizability has also been exploited [50] to construct other classes of potentials.

5.7   Phase Equivalent Potentials

An early work on phase equivalent potentials is due to Bargmann [51] who solved linear, quadratic, exponentially decreasing and rational potentials within a phase equivalent system. With the advent of SUSYQM, Sukumar [52] utilized the factorization scheme to study phase-shift differences and partner supersymmetric Hamiltonians. Subsequently, Baye [53] showed that a pair of phase equivalent potentials could be generated employing two successive supersymmetric transformations with the potentials supporting different number of bound states. Later, general analytic expressions were obtained [54] which express suppression of the N lowest bound states of the spectrum.

When the procedure of factorizability is used to modify the bound spectrum, the phase shifts are also modified because of Levinson theorem [55-57]. The latter states [51,58] that two phase equivalent potentials are identical if both fall off sufficiently and rapidly at large distances and if neither yields a bound state. In the case of an iterative supersymmetric procedure, since the number of bound states vary, the singularity of the phase equivalent potentials can also change.

During recent times the formalism of developing phase equivalent potentials has been expanded to include arbitrary modifications of the energy spectrum. The works include [59-61] the one of Amado [59] who explored a class of exactly solvable one-dimensional problems and Levai, Baye, and Sparenberg [60] who extended phase equivalence to the generalized Ginocchio potentials and were successful in obtaining closed algebraic expressions for the phase equivalent partners. It may be mentioned that Roychoudhury and his collaboraters [62] also made an extensive study of the generation of isospectral Hamiltonians to construct new potentials (some of which are phase equivalent) from a given starting potential.

In the following we demonstrate how phase equivalent potentials can be derived using the techniques of SUSY.

Let us consider a radial ($r > 0$) Hamiltonian $H_+^{(1)}$ factorized according to

$$H_+^{(1)} = -\frac{1}{2}\frac{d^2}{dr^2} + V_+^{(1)}(r) \tag{5.85}$$

$$= \frac{1}{2}A_1^+ A_1 + E^{(1)} \tag{5.86}$$

where $V_+^{(1)}(r)$ is a given potential and $E^{(1)}$ is some arbitrary negative energy. Analogous to (2.86) we take

$$A_1 = \frac{d}{dr} + W(r)$$

$$A_1^+ = -\frac{d}{dr} + W(r) \tag{5.87}$$

By reversing the factors in (5.86) we can at once write down the supersymmetric partner to $H_+^{(1)}$, namely $H_-^{(1)}$, which reads

$$H_-^{(1)} = \frac{1}{2}A_1 A_1^+ + E^{(1)} \tag{5.88}$$

$$= -\frac{1}{2}\frac{d^2}{dr^2} + V_-^{(1)}(r) \tag{5.89}$$

Note that the spectrum of $H_-^{(1)}$ is almost identical to $H_+^{(1)}$ with the possible exception of $E^{(1)}$.

Using (2.84) it follows that

$$V_-^{(1)}(r) = V_+^{(1)}(r) - \frac{d^2}{dr^2}\log\psi_0^{(1)}(r) \tag{5.90}$$

where $\psi_0^{(1)}(r)$ is the accompanying solution to $E^{(1)}$ of the Schroedinger equation (5.85). Moreover the superpotential $W(r)$ is expressible in terms of $\psi_0^{(1)}$

$$W(r) = -\frac{[\psi_0^{(1)}]'}{\psi_0^{(1)}} \tag{5.91}$$

Expressions (5.86) and (5.88) constitute what may be called a "first stage" factorization. However, as pointed out in the previous section the factorizability of the Schroedinger Hamiltonian is not unique.

Indeed we can consider a "second stage" factorization induced by the pairs

$$A_2 = \frac{d}{dr} + W(r) + \chi(r)$$

$$A_2^+ = -\frac{d}{dr} + W(r) + \chi(r) \tag{5.92}$$

where $\chi(r)$ is given by [see (5.80) along with (5.91)]

$$\chi(r) = \frac{e^{-2 \int^r W(t)dt}}{\beta + \int^r e^{-2 \int^r W(t)dt} dt} \tag{5.93}$$

with $\beta \in \mathbb{R}$.

The factors A_2 and A_2^+ give rise to a new Hamiltonian which we denote as $H_+^{(2)}$

$$H_+^{(2)} = -\frac{1}{2}\frac{d^2}{dr^2} + V_+^{(2)}(r) \tag{5.94}$$

$$= \frac{1}{2}A_2^+ A_2 + E^{(1)} \tag{5.95}$$

$H_+^{(2)}$ has the partner $H_-^{(2)}$ namely

$$H_-^{(2)} = -\frac{1}{2}\frac{d^2}{dr^2} + V_-^{(2)}(r) \tag{5.96}$$

$$= \frac{1}{2}A_2 A_2^+ + E^{(1)} \tag{5.97}$$

A little calculation shows that $V_-^{(2)}(r)$ can be put in the form

$$V_2 \equiv V_-^{(2)}(r) = V_+^{(1)}(r) - \frac{d^2}{dr^2} \log\left[\beta + \int_r^\infty e^{-2\int W(t)dt} dt\right] \tag{5.98}$$

The potential V_2, which has no singularity at finite distances, is phase equivalent to $V_+^{(1)}(r)$ with the following options for β namely, $\beta = -1, \alpha$ or $\alpha(1-\alpha)^{-1}$ with $\alpha > 0$ being arbitrary. While for the first two choices of β, $E^{(1)}$ is physical for $H_+^{(1)}$, that for the second one, $E^{(1)}$ is nonphysical for $H_+^{(1)}$. Physically this means [54] that for $\beta = -1$, a suppressed bound state continues to remain suppressed after two successive factorizations; for $\beta = \alpha$, a new bound state appears at $E^{(1)}$ along with a parameter in the potential; for $\beta = \alpha(1-\alpha)^{-1}$ the bound spectrum remains unchanged but at the cost of introducing a parameter in the potential. Note that the third case can also be looked upon as a combination of the possibilities (a) and (b).

A few remarks on the wave functions of $H_-^{(1)}$ and $H_-^{(2)}$ are in order

(i) The solution $\psi_-^{(1)}(r)$ of $H_-^{(1)}$ can be given in terms of $\psi_0^{(1)}(r)$ and the solution $\psi_0(r)$ of a reference Hamiltonian $H_0 = -\frac{d^2}{dr^2} + V_0(r)$ as

$$\psi_-^{(1)}(r) = \left[\psi_0^{(1)}\right]^{-1} \int_r^\infty \psi_0^{(1)} \psi_0 dt \qquad (5.99)$$

One allows $V_0(r)$ to be singular at the origin which, excluding Coulomb and centrifugal parts, looks like

$$V_0(r) \simeq \frac{n(n+1)}{r^2} \qquad (5.100)$$

where n is nonnegative and not necessary to be identified [56] with the orbital momenta l.

Note that ψ_0 corresponds to H_0 for some arbitrary energy $E(\neq E^{(1)})$ which is bounded at infinity and factorizations in (5.86) are carried out corresponding to the set of solutions $[\psi_0^{(1)}, E^{(1)}]$ of H_0. It is clear from (5.99) that there is a modification of phase shifts.

(ii) On the other hand, the solution of $H_-^{(2)}$ reads for $E \neq E^{(1)}$

$$\psi_-^{(2)}(r) = \psi_0 - \psi_0^{(1)} \frac{\int_r^\infty \psi_0^{(1)} \psi_0 dt}{\beta + \int_0^\infty [\psi_0^{(1)}]^2 dt} \qquad (5.101)$$

showing that $\psi_-^{(2)}$ and ψ_0 are different by a short-ranged term. With (5.101) there is no modification of the phase shifts as a result V_2 is phase equivalent to V_0. Note that (5.101) is valid at all energy values corresponding to both physical and nonphysical solutions of $H_-^{(2)}$.

It may be pointed out that when $E = E^{(1)}$ the corresponding solution of $H_-^{(1)}$ does not vanish at the origin indicating suppression of the bound state. In the case of $\psi_-^{(2)}(r)$, for $E = E^{(1)}$, there is a modification of the expression (5.101) by a normalization factor.

We now illustrate the procedure of deriving $V_2(r)$ from a given superpotential. We consider the case of Bargmann potential.

Bargmann potential is a rational potential of the type [51]

$$V^B(r) = 3(r - \alpha) \left[\frac{(r - \alpha)^3 - 2\gamma^3}{\omega^2(r)} \right] \tag{5.102}$$

where α and γ are the parameters of the potential and $\omega(r) = (r - \alpha)^3 + \gamma^3$.

The interest in $V^B(r)$ comes from the fact that the potential

$$V(r) = -Am^2 \frac{e^{-mr}}{(1 + Ae^{-mr})^2}, \quad m > 0 \tag{5.103}$$

introduced by Eckart [65], is well known to be phase equivalent to (5.102) for a certain choice of the parameters A and m. It is clear from (5.103) that for $A < 0$, the potential $V(r)$ is repulsive and has no bound state.

If $V^B(r)$ is expressed as $\frac{1}{2}(W^2 - W')$, the superpotential $W(r)$ is readily obtained as [66]

$$W(r) = \frac{3(r - \alpha)^2}{\omega} - \frac{1}{r - \alpha} \tag{5.104}$$

From (5.98) we then derive

$$
\begin{aligned}
V_2^B &= 3(r - \alpha) \left[\frac{(r - \alpha)^3 - 2\gamma^3}{\omega^2} \right] \\
&+ \frac{6(r - \alpha)}{3\beta\omega^2 + \omega} \\
&- \frac{9(r - \alpha)^4 \left[6(r - \alpha)^3 + 6\gamma^3 + 1 \right]}{(3\beta\omega^2 + \omega)^2}
\end{aligned} \tag{5.105}
$$

It may be remarked that for $V^B(r)$ there is a stationary state of zero energy when $\alpha = 0$. On the other hand, for $\alpha > 0$ the only bound state is that of a negative energy.

Finally, (5.98) can be extended to the most general form by considering a set of different but arbitrary negative energies. However, even for the simplest examples extracting an analytical form of successive phase equivalent potentials is extremely difficult. Indeed one often has to resort to computer calculations [57] to obtain the necessary expressions. From a practical point of view supersymmetric transformations have been exploited to have phase equivalent removal of the forbidden states of a deep potential thus leading to a shallow potential. In this context other types of potentials have been studied such as complex (optical) potentials, potentials having a linear dependence on energy, and those of coupled-channel types [63,64]. Physical properties of deep and shallow phase equivalent potentials encountered in nuclear physics [67-69] have also been compared with.

5.8 Generation of Exactly Solvable Potentials in SUSYQM

Determination [6,70-82] of exactly solvable potentials found an impetus chiefly through the works of Bhattacharjie and Sudarshan [71,72] and also Natanzon [74] who derived general properties of the potentials for which the Schroedinger equation could be solved by means of hypergeometric, confluent hypergeometric, and Bessel functions. In this connection mention should be made of the work of Ginocchio [75] who also studied potentials that are finite and symmetric about the origin and expressible in terms of Gegenbour polynomials. Of course Ginocchio potentials belong to a sublass of Natanzon's.

    Let us now take a quick look at some of the potentials which can be generated in a natural way by employing a change of variables in a given Schroedinger equation. In this regard we consider a mapping $x \to g(x)$ which transforms the Schroedinger equation

$$\left[-\frac{1}{2}\frac{d^2}{dx^2} + \{V(x) - E\} \right] \psi(x) = 0 \qquad (5.106)$$

into a hypergeometric form. The potential can be presented as

$$V(g(x)) = \frac{c_0 g(g-1) + c_1(1-g) + c_2 g}{R(g)} - \frac{1}{2}\{g, x\} \qquad (5.107)$$

where $R(g)$ is

$$R(g) = A_i(g - g_i)^2 + B_i(g - g_i) + C_i, \quad g_i = 0, 1 \qquad (5.108)$$

and the Schwartzian derivative is defined by

$$\{g, x\} = \frac{g'''}{g'} - \frac{3}{2}\left(\frac{g''}{g'}\right)^2 \qquad (5.109a)$$

In (5.107) and (5.108), c_0, c_1, c_2, A_i, B_i, and C_i appear as constants. The transformation $g(x)$ is obtained from the differential equation

$$(g')^2 = \frac{4g^2(1-g)^2}{R(g)} \tag{5.109b}$$

In (5.109) the primes denote derivatives with respect to x (5.107) and constitute Natanzon class of potentials.

As can be easily seen, the following simple choices of $R(g)$ yield some of the already known potentials

$$
\left\{
\begin{aligned}
R(g) &= \frac{g^2}{a^2} : g = 1 - \exp[-2a(x-x_0)], \\
V(x) &= \frac{a^2}{2} \left[(c_0 - c_2)\left\{ 1 - \coth a(x-x_0) \right\} \right. \\
& \qquad \left. + \tfrac{c_1}{4} \operatorname{cosech}^2 a(x-x_0) \right]
\end{aligned}
\right.
$$

$$
\left\{
\begin{aligned}
R(g) &= \frac{g}{a^2} : g = \tanh^2[a(x-x_0)], \\
V(x) &= \frac{a^2}{2} \left[\left(c_1 + \frac{3}{4} \right) \operatorname{cosech}^2 a(x-x_0) \right. \\
& \qquad \left. - \left(c_0 + \tfrac{3}{4} \right) \operatorname{sech}^2 a(x-x_0) \right]
\end{aligned}
\right.
$$

$$
\left\{
\begin{aligned}
R(g) &= \frac{1}{a^2} : g = \frac{\exp[2a(x-x_0)]}{1 + \exp[2a(x-x_0)]}, \\
V(x) &= \frac{a^2}{2} \left[\frac{1}{2}(c_1 - c_2)\left\{ 1 - \tanh a(x-x_0) \right\} \right. \\
& \qquad \left. - \tfrac{1}{4} c_0 \operatorname{sech}^2 a(x-x_0) \right]
\end{aligned}
\right. \tag{5.110a, b, c}
$$

The potentials (5.110a), (5.110b) and (5.110c) can be recognized to be the Eckart I, Poschl-Teller II, and Rosen-Morse I, respectively. The corresponding wave functions can be expressed in terms of Jacobi polynomials which in turn are known in terms of hypergeometric functions. As we know from the results of Section 5.3 and Table 5.1, these potentials are SI in nature. So SI potentials are contained in the Natanzon class of potentials.

Searching for special functions which are solutions of the Schroedinger equation has proven to be a useful procedure to identify solvable potentials. Within SUSYQM this approach has helped explore not only the SI potentials but also shape-noninvariant ones [6].

Even potentials derived from other schemes [83-87] have been found to obey the Schroedinger equation whose solutions are governed by typical special functions [88]. In the following however we would be interested in SI potentials only.

Let us impose a transformation $\psi = f(x)F(g(x))$ on the Schroedinger equation (5.106) to derive a very general form of a second-order homogeneous linear differential equation namely

$$\frac{d^2 F}{dg^2} + Q(g)\frac{dF}{dg} + R(g)F(g) = 0 \tag{5.111}$$

where the function $Q(g)$ and $R(g)$ are given by

$$Q(g) = \frac{g''}{(g')^2} + \frac{2f'}{fg'} \tag{5.112}$$

$$R(g) = \frac{f''}{f(g')^2} + 2\frac{E - V(x)}{(g')^2} \tag{5.113}$$

In the above primes denote derivatives with respect to x.

The form (5.111) enables us to touch those differential equations which are well-defined for any particular class of special functions. Such differential equations offer explicit expressions for $Q(g)$ and $R(g)$ which can then be trialed for various plausible choices of $g(x)$ leading to the determination of exactly solvable potentials. Orthogonal polynomials in general have the virtue that the conditions of the partner potentials in SUSYQM appear in a particular way and are met by them.

Using the trivial equality $\frac{f''}{f} = \left(\frac{f'}{f}\right)' + \left(\frac{f'}{f}\right)^2$ we may express (5.113) as

$$2[E - V(x)] = Rg'^2 - \left[\left(\frac{f'}{f}\right)^2 + \left(\frac{f'}{f}\right)'\right] \tag{5.114}$$

Eliminating now f'/f from (5.112) and (5.114) we obtain

$$2[E - V(x)] = \frac{1}{2}\{g, x\} + \left[R(g) - \frac{1}{2}\frac{dQ}{dg} - \frac{1}{4}Q^2\right](g')^2 \tag{5.115}$$

Equation (5.115) is the key equation to be explored. The main point is that if a suitable g is found which makes at least one term in the right-hand-side of (5.115) reduced to a constant, it can be immediately identified with the energy E and the remaining terms make up for the potential energy. Since $Q(g)$ and $R(g)$ are known beforehand we should identify (5.111) with a particular differential equation with known special functions as solutions [89-92]; all this actually amounts to experimenting with different choices of $g(x)$ to guess at a reasonable form of the potential. Of course, often a transformation of parameters may be necessitated, as the following example will clarify, to lump the entire n dependence to the constant term which can then be interpreted to stand for the energy levels. It is worth remarking that the present methodology [80] of generating potentials encompasses not only Bhattacharjie and Sudarshan but also Natanzon schemes.

To view (5.115) in a supersymmetric perspective we observe that whenever $R(g) = 0$ holds we are led to a correspondence

$$V(x) - E = \frac{1}{2}\left[\left(\frac{f'}{f}\right)^2 + \left(\frac{f'}{f}\right)'\right] = \frac{1}{2}\left(W^2 - W'\right) \equiv H_+ \quad (5.116)$$

from (5.114). In (5.116) W has been defined as

$$W = -\frac{f'}{f} = -(\log f)' \qquad (5.117)$$

For Jacobi $\left[P_n^{i,j}(g)\right]$ and Laguerre $[L_n^i(g)]$ polynomials one have

$$\left[P_n^{i,j}(g)\right] : Q(g) = \frac{j-i}{1-g^2} - (i+j+2)\frac{g}{1-g^2}$$

$$R(g) = \frac{n(n+i+j+1)}{1-g^2} \qquad (5.118)$$

$$L_n^i(g) : Q(g) = \frac{i-g+1}{g}$$

$$R(g) = \frac{n}{g} \qquad (5.119)$$

So it can be seen that $R(g) = 0$ for the value $n = 0$. Thus E in (5.116) corresponds to $n = 0$. Note however that the Bessel equation does not fulfill the criterion of $R(g) = 0$ for $n = 0$.

To inquire into the functioning of the above methodology let us analyze a particular case first. Identifying the Schroedinger wave function ψ with a confluent hypergeometric function $F(-n, \beta, g)$ and writing g as $g(x) = \rho h(x)$, ρ a constant, we have from (5.115)

$$2\left[E_n - V(x)\right] = \frac{1}{2}\{h, x\} + \rho\frac{(h')^2}{h}\left(n + \frac{\beta}{2}\right)$$
$$-\frac{\rho^2}{4}(h')^2 + \left(\frac{h'}{h}\right)^2\frac{\beta}{2}\left(1 - \frac{\beta}{2}\right) \quad (5.120)$$

where $F(a, c, g)$ satisfies the differential equation $\frac{d^2 F}{dg^2} + \frac{\{c-g\}}{g}\frac{dF}{dg} - \frac{a}{g}F = 0$.

Since we need at least one constant term in the right-hand-side of (5.120) to match with E in the left-hand-side, we have following options: either we set $\frac{h'^2}{h} = c$ or $h'^2 = c$ or $\frac{h'^2}{h^2} = c$, c being a constant. To examine a specific case [93], let us take the second one which implies $h = \sqrt{c}x$. From (5.120) we are led to

$$2\left[E_n - V(x)\right] = -\frac{c\rho^2}{4} + \frac{\rho\sqrt{c}}{x}\left(n + \frac{\beta}{2}\right) + \frac{1}{x^2}\frac{\beta}{2}\left(1 - \frac{\beta}{2}\right) \quad (5.121)$$

However, in the right-hand-side of the above equation the second term is both x and n dependent. So to be a truly unambiguous potential which is free from the presence of n, we have to get rid of the dependence of n it. Note that the n index of the confluent hypergeometric function is made to play the role of the quantum number for the energy levels in the left-hand-side (5.121). We set $\rho_n = A\left(n + \frac{\beta}{2}\right)^{-1}$ which allows us to rewrite (5.121) as

$$2\left[E_n - V(x)\right] = \frac{A\sqrt{c}}{x} + \frac{1}{x^2}\frac{\beta}{2}\left(1 - \frac{\beta}{2}\right) - \frac{c}{4}\rho_n^2 \quad (5.122)$$

In this way the n dependence has been shifted entirely into the constant term which can now be regarded as the energy variable. Identifying β as $2(l + 1)$, $A\sqrt{c}$ as 2 and restricting to the half-line $(0, \infty)$ we find that (5.122) conforms to the hydrogen atom problem with $V(r) = -\frac{1}{r} + \frac{l(l+1)}{2r^2}$ where the parameters $\hbar, m, e$, and Z have been scaled to unity because of $A\sqrt{c} = 2$. The Coulomb problem is certainly SI, the relevant parameters being $c_0 = l$ and $c_1 = l + 1$, l being the principal quantum number.

In connection with SI potentials in SUSYQM, Levai [80] in a series of papers has made a systematic analysis of the basic equation

(5.115). Applying it to the Jacobi, generalised Laguerre and Hermite polynomials, he has been led to several families of secondary differential equations. Their solutions reveal the existence of 12 different SI potentials [88] with the scope of finding new ones quite remote. Levai's classification scheme may be summarised in terms of six classes as shown in Table 5.2. Note that the orthogonal polynomials like Gegenbauer, Chebyshev, and Legendre have not been considered since these are expressible as special cases from $P_n^{i,j}(g)$.

5.9 Conditionally Solvable Potentials and SUSY

Interest in conditionally exactly solvable (CES) systems has been motivated by the fact that in quantum mechanics exactly solvable potentials are hard to come by. CES systems are those for which the energy spectra is known under certain constraint conditions among the potential parameters.

CES potentials can be obtained [94] from the secondary differential equation (5.111) by putting $Q(g) = 0$. It implies

$$g'f^2 = \text{constant},$$
$$\psi = (g')^{-1/2} F[g(x)] \qquad (5.123)$$

and from (5.115)

$$2[E - V(x)] = \frac{1}{2}\{g, x\} + R(g')^2 \qquad (5.124)$$

To exploit (5.124) let us set $R = 2[E_T - V_T(g)]$ and use the transformation $x = f(g)$. We get the result

$$V_T(g) - E_T = [f'(g)]^2 [V\{f(g)\} - E] + \frac{1}{2}\Delta V(g) \qquad (5.125)$$

where

$$\Delta V(g) = \left[-\frac{1}{2}\frac{f'''(g)}{f'(g)} + \frac{3}{4}\left(\frac{f''(g)}{f'(g)}\right)^2 \right] \qquad (5.126)$$

In the above equations the primes stand for differentiations with respect to the variable g.

Table 5.2

A list of 12 different SI potentials for different choices of g. Here m and $\hbar$ are not explicitly displayed. The results in this Table are consistent with the list provided in [80].

Class	Differential Eqn.	Solution for g	Wave Function	Remarks
I	$\frac{g'^2}{1-g^2} = C$	(i) $i\sinh(\alpha x)$	$(1-g^2)^{-\nu/\omega}\times \exp\{-\lambda\tan^{-1}(-ig)\}P_n^{i,j}(g)$	$i = -\sqrt{-1}\lambda - \nu - \frac{1}{2}$, $j = \sqrt{-1}\lambda - \nu - \frac{1}{2}$
	$\left(\nu = \frac{A}{\alpha}, \lambda = \frac{\beta}{\alpha}\right)$	(ii) $\cosh(\alpha x)$	$(g-1)^{\frac{\lambda-\nu}{2}}(g+1)^{-(\lambda-\nu)/2}\times P_n^{i,j}(g)$	$i = \lambda - \nu - \frac{1}{2}$, $j = -\nu - \lambda - \frac{1}{2}$
		(iii) $\cos(\alpha x)$	$(1-g)^{\frac{\nu-\lambda}{2}}(1+g)^{(\nu+\lambda)/2}\times P_n^{i,j}(g)$	$i = \nu - \lambda - \frac{1}{2}$, $j = \nu + \lambda - \frac{1}{2}$
		(iv) $\cos(2\alpha x)$	$(1-g)^{\frac{\lambda}{2}}(1+g)^{\nu/2}\times P_n^{i,j}(g)$	$i = \lambda - \frac{1}{2}$, $j = \nu - \frac{1}{2}$
		(v) $\cosh(2\alpha x)$	$(g-1)^{\frac{\lambda}{2}}(g+1)^{-\nu/2}\times P_n^{i,j}(g)$	$i = \lambda - \frac{1}{2}$, $j = -\nu - \frac{1}{2}$
II	$\frac{g'^2}{1-g^2} = C$	(i) $i\tanh(\alpha x)$	$(1-g)^{\frac{\nu-n+\mu_-}{2}}\times (1+g)^{\frac{\nu-n-\mu_-}{2}}P_n^{i,j}(g)$	$i = -\nu - n + \mu_-$, $j = -\nu - n - \mu_-$
$\mu_\pm = \frac{\lambda}{\nu\pm m}$	$\left(\lambda = \frac{\beta}{\alpha^2}\right)$	(ii) $\cot h(\alpha x)$	$(g-1)^{-(\nu+n-\mu_+)/2}\times (g+1)^{-(\nu+n+\mu_+)/2}P_n^{i,j}(g)$	$i = \nu - n + \mu_+$, $j = -\nu - n - \mu_+$
		(iii) $-i\cot(\alpha x)$	$(g^2-1)^{(\nu-n)/2}\exp(\mu-\alpha x)\times$ $i = \nu - n + \sqrt{-1}\mu_-,$ $P_n^{i,j}(g)$	$j = \nu - n - \sqrt{-1}\mu_-$
III	$\frac{(g')^2}{g} = C$	$\frac{1}{2}\omega x^2$	$g^{(l+1)/2}\exp(-\frac{g}{2})L_n^{(l+1)/2}(g)$	—
IV	$(g')^2 = C$	$\frac{e^2 x}{n+l+1}$	$g^{(l+1)/2}\exp(-\frac{g}{2})L_n^{(2l+1)/2}(g)$	—
V	$\frac{(g')^2}{g} = C$	$\frac{2\beta}{\alpha\exp(-a\omega)}$	$g^{(\nu-n)/2}\exp(-\frac{g}{2})L_n^{(2\nu-2n)}(g)$	—
VI	$(g')^2 = C$	$\left(\frac{1}{2}\omega\right)^{1/2}\left(n - \frac{2b}{\omega}\right)$	$\exp(-\frac{1}{2}g^2)H_n(g)$	—

Table 5.2 (continued)

Solution for g	Potentials	Energy levels
(i) $i\sinh(\alpha x)$	$A^2 + (B^2 - A^2 - A\alpha)\sec h^2(\alpha x)$ $+ B(2A+\alpha)\sec h(\alpha x)\tanh(\alpha x)$	$A^2 - (A - n\alpha)^2$
(ii) $\cosh(\alpha x)$	$A^2 + (A^2 + B^2 + A\alpha)\mathrm{cosech}^2(\alpha x)$ $+ B(2A+\alpha)\mathrm{cosech}(\alpha x)\coth(\alpha x)$	$A^2 - (A - n\alpha)^2$
(iii) $\cos(\alpha x)$	$-A^2 + (A^2 + B^2 - A\alpha)\mathrm{cosec}^2(\alpha x)$ $+ B(B-\alpha)\mathrm{cosec}^2(\alpha x)$	$(A + n\alpha)^2 - A^2$ $-(A+B)^2$
(iv) $\cos(2\alpha x)$	$-(A+B)^2 + A(A-\alpha)\sec^2(\alpha x)$ $+ B(B-\alpha)\mathrm{cosec}^2(\alpha x)$	$(A+B+2n\alpha)^2$ $-(A+B)^2$
(v) $\cosh(2\alpha x)$	$(A-B)^2 - A(A+\alpha)\sec^2(\alpha x)$ $+ B(B-\alpha)\mathrm{cosech}^2(\alpha x)$	$(A-B)^2$ $-(A-B-2n\alpha)^2$
(i) $\tanh(\alpha x)$	$A^2 + \frac{B^2}{A^2} - A(A+\alpha)\sec h^2(\alpha x)$ $+ 2B\tanh(\alpha x)$	$A^2 + \frac{B^2}{A^2} - (A+n\alpha)^2$ $-\frac{B^2}{(A+n\alpha)^2}$
(ii) $\coth(\alpha x)$	$A^2 + \frac{B^2}{A^2} + A(A-\alpha)\mathrm{cosech}^2(\alpha x)$ $+ 2B\coth(\alpha x)$	$A^2 + \frac{B^2}{A^2} - (A+n\alpha)^2$ $-\frac{B^2}{(A+n\alpha)^2}$
(iii) $-i\coth(\alpha x)$	$-A^2 + \frac{B^2}{A^2} + A(A+\alpha)\mathrm{cosec}^2(\alpha x)$ $- 2B\cot(\alpha x)$	$-A^2 + \frac{B^2}{A^2} + (A-n\alpha)^2$ $-\frac{B^2}{(A-n\alpha)^2}$
$\frac{1}{2}\omega x^2$	$\frac{1}{4}\omega^2 x^2 + l(l+1)/x^2 - (l+\frac{3}{2})\omega$	$2n\omega$
$\frac{e^2 x}{n+l+1}$	$\frac{1}{4}\frac{e^4}{(l+1)^2} - \frac{e^2}{x} + \frac{l(l+1)}{x^2}$	$\frac{1}{4}\frac{e^2}{(l+1)^2} - \frac{1}{4}\frac{e^2}{(n+l+1)^2}$
$\alpha\exp(-\alpha x)$	$A^2 - B(2A+\alpha)\exp(-\alpha x)$ $+ B^2\exp(-2\alpha x)$	$A^2 - (A - n\alpha)^2$
$\left(\frac{1}{2}\omega\right)^{1/2}\left(n - \frac{2b}{\omega}\right)$	$\frac{1}{2}\omega + \frac{1}{4}\omega^2 x^2$	ωm

In recent times the results (5.125) and (5.126) have been used [95–101] to get CES potentials in the form $V[f(g)]$ by a judicious choice of the transformation function $f(g)$ and assigning to V_T a convenient exactly solvable potential whose energy spectrum and eigenfunctions are known. Let us now consider a few examples regarding the applicability of the results (5.125) and (5.126). We restrict to those which exhibit SUSY [97].

Choose the mapping function to be

$$x = f(g) = \log(\sinh g) \tag{5.127}$$

it is obvious that the domain of the variable g is the half-line $(0, \infty)$ for $x \in (-\infty, \infty)$. The quantity $\Delta V(g)$ becomes

$$\Delta V(g) = \frac{3}{4} \tanh^2 g - \frac{1}{4} \operatorname{cosech}^2 g - \frac{3}{4} \tag{5.128}$$

At this stage it becomes imperative to choose a particular form for $V[f(g)]$. This gives $V_T(g)$ from (5.125). If the properties of $V_T(g)$ are known then those of $V[f(g)]$ can be easily determined. Let us have

$$V[f(g)] = \frac{1}{2} \left(a \tanh^2 g - b \tanh g + c \tanh^4 g \right) \tag{5.129}$$

Then it easily follows from (5.125) that for $c = -3/4$

$$V'_T \equiv 2V_T = -b \coth g - \left(E_n + \frac{1}{4} \right) \operatorname{cosech}^2 g \tag{5.130}$$

$$(E_T)_n = -a + E_n + \frac{3}{4} \tag{5.131}$$

where we have taken

$$E \equiv \frac{1}{2} E_n$$

$$E_T \equiv \frac{1}{2} (E_T)_n \tag{5.132}$$

Comparison with Table 5.1 reflects that V'_T is essentially the Eckart I potential

$$V(r) = -2B \coth r + A(A-1) \operatorname{cosech}^2 r \tag{5.133}$$

whose energy spectrum is

$$E_n = -(A+n)^2 - \frac{B^2}{(A+n)^2} \tag{5.134}$$

Moreover the corresponding eigenfunctions of (5.133) are known in terms of the Jacobi polynomials. Comparing $V(r)$ with $V_T'(g)$ we find that g plays the role of r with $b = 2B$ and $E_n = A(1-A) - \frac{1}{4}$. The latter implies $A = \frac{1}{2} + \sqrt{-E}$. On the other hand, if we compare (5.131) with (5.134) it follows that

$$a - E_n - \frac{3}{4} = \left[n + \frac{1}{2} + \sqrt{-E_n} \right]^2$$
$$+ \frac{b^2}{4 \left(n + \frac{1}{2} + \sqrt{-E_n} \right)^2} \tag{5.135}$$

Writing $E_n = -\epsilon_n$ to have A real, (5.135) can be expressed in the form of a cubic equation for the quantity $\sqrt{\epsilon_n}$. It turns out that [97] of the three roots, two can be discarded requiring consistency with the potential for $b = 0$. We also get ψ from (5.123) since, as already remarked, the eigenfunctions of (5.133) are available in terms of the Jacobi polynomials.

In terms of x we thus have a CES potential from (5.129)

$$V(x) = \frac{1}{2} \left[\frac{a}{1 + e^{-2x}} - \frac{b}{(1 + e^{-2x})^{1/2}} - \frac{3}{4(1 + e^{-2x})^2} \right] \tag{5.136}$$

whose parameters a and b are constrained by (5.135).

Interestingly we can also write down a superpotential $W(x)$ for (5.136) which reads

$$W(x) = \frac{p}{(1 + e^{-2x})^{1/2}} - \frac{1}{2(1 + e^{-2x})} - \sqrt{\epsilon_0} \tag{5.137}$$

where $p = b(1 + 2\sqrt{\epsilon_0})^{-1/2}$ and (5.135) has been used. In principle one can encompass the CES potentials by obtaining the partner potential $\frac{1}{2}(W^2 + W')$.

Similarly, we may consider another type of $V[f(g)]$ given by

$$V[f(g)] = \frac{1}{2} \left(a \tanh^2 g - b \operatorname{sech} g - \frac{3}{4} \tanh^4 g \right) \tag{5.138}$$

which will induce

$$V'_T = 2V_T = -b \cosech g \coth g - \left(E_n + \frac{1}{4}\right) \cosech^2 g \quad (5.139)$$

Since from Table 5.1, (5.139) corresponds to the Rosen-Morse II potential, its associated eigenfunctions and energy levels are known. As before, these determine the corresponding eigenfunctions and eigenvalues of (5.138). Thus we have another CES potential given by

$$V(x) = \frac{a}{1 + e^{-2x}} - \frac{be^{-x}}{(1 + e^{-2x})^{1/2}} - \frac{3}{4(1 + e^{-2x})^2} \quad (5.140)$$

whose parameters a and b are restricted by

$$\left(\epsilon_n + a - \frac{3}{4}\right)^{1/2} = \frac{1}{2}\left[(\epsilon_n + b)^{1/2} - (\epsilon_n - b)^{1/2}\right] - \left(n + \frac{1}{2}\right) \quad (5.141)$$

The potential (5.139) also affords a superpotential $W(x)$ in the form

$$W(x) = a + \frac{1}{2(1 + e^{2x})} - \frac{b}{(1 + e^{2x})^{1/2}} \quad (5.142)$$

Other examples of CES potentials include the singular one [96]

$$V(x) = \frac{a}{r} + \frac{b}{r^{1/2}} - \frac{3}{32r^2} \quad (5.143)$$

It turns out that the restriction required on the parameters a and b for (5.143) to be a CES potential is the same as the one for which this potential can be put in a supersymmetric form [98]. The list of CES potentials has been expanded to include Natanzan potentials also [99,100]. By fixing the free parameters they give rise to fractional power, long-range and strongly anharmonic terms. Apart from SUSY, CES potentials have also appeared as ordinary quantum mechanical problems as well. One such instance is a class of partially solvable rational potentials with known zero-energy solutions. For some of them the zero-energy wave function has been found to be normalizable and to describe a bound state [94].

5.10   References

[1] F. Cooper, A. Khare, and U. Sukhatme, *Phys. Rep.*, **251**, 267, 1995.

[2] C.V. Sukumar, *J. Phys. A. Math. Gen.*, **18**, L57, 1985.

[3] C.V. Sukumar, *J. Phys. A. Math. Gen.*, **18**, 2917, 1985.

[4] L.E. Gendenshtein, *JETP Lett.*, **38**, 356, 1983.

[5] R. Dutt, A. Khare, and U.P. Sukhatme, *Am. J. Phys.*, **56**, 163, 1987.

[6] F. Cooper, J.N. Ginocchio, and A. Khare, *Phys. Rev.*, **D36**, 2458, 1987.

[7] L. Infeld and T.E. Hull, *Rev. Mod. Phys.*, **23**, 21, 1951.

[8] L. Infeld and T.E. Hull, *Phys. Rev.*, **74**, 905, 1948.

[9] D. L. Pursey, *Phys. Rev.*, **D33**, 1048, 1986.

[10] D. L. Pursey, *Phys. Rev.*, **D33**, 2267, 1986.

[11] M. Luban and D.L. Pursey, *Phys. Rev.*, **D33**, 431, 1986.

[12] R. Montemayor, *Phys. Rev.*, **A36**, 1562, 1987.

[13] A. Stahlhofen and K. Bleuler Nuovo Cim, **B104**, 447, 1989.

[14] V.G. Bagrov and B.F. Samsonov, *Theor. Math. Phys.*, **104**, 1051, 1945.

[15] L.J. Boya, *Eur. J. Phys.*, **9**, 139, 1988.

[16] R. Montemayor and L.D. Salem, *Phys. Rev.*, **A40**, 2170, 1989.

[17] E. Schroedinger, *Proc. Roy Irish Acad.*, **A46**, 9, 1940.

[18] E. Schroedinger, *Proc. Roy Irish Acad.*, **A46**, 183, 1941.

[19] E. Schroedinger, *Proc. Roy Irish Acad.*, **A47**, 53, 1941.

[20] H. Weyl, *The Theory of Groups and Quantum Mechanics*, E.P. Dutton and Co., New York, 1931.

[21] P.A.M. Dirac, *Principles of Quantum Mechanics*, 2nd ed., Clarendon Press, Oxford, 1947.

[22] A.F.C. Stevenson, *Phys. Rev.*, **59**, 842, 1941.

[23] A. Lahiri, P.K. Roy, and B. Bagchi, *Int. J. Mod. Phys.*, **A5**, 1383, 1990.

[24] A. Khare and U.P. Sukhatme, *J. Phys. A. Math. Gen.*, **26**, L901, 1993.

[25] A.B. Shabat and R.I. Yamilov, *Leningrad Math. J.*, **2**, 377, 1991.

[26] V. De. Alfaro, S. Fubini, and G. Furlan, *Nuovo Cim.*, **A34**, 569, 1976.

[27] A.P. Veslov and A.B. Shabat, *Funct. Anal. Appl.*, **27**, n2,1, 1993.

[28] A. Shabat, *Inverse Prob.*, **8**, 303, 1992.

[29] V. Spiridonov, *Mod. Phys. Lett.*, **A7**, 1241, 1992.

[30] V. Spiridonov, *Comm. Theor. Phys.*, **2**, 149, 1993.

[31] V. Spiridonov, *Phys. Rev. Lett.*, **69**, 398, 1992.

[32] T. Fukui, *Phys. Lett.*, **A189**, 7, 1994.

[33] A.B. Balantekin, M.A. Cândido Riberio, and A.N.F. Aleixo, *J. Phys. A. Math. Gen.*, **32**, 2785, 1999.

[34] E.P. Wigner, *Phys. Rev.*, **77**, 711, 1950.

[35] N. Mukunda, E.C.G. Sudarshan, J. Sharma, and C.L. Mehta, *J. Math. Phys.*, **21**, 2386, 1980.

[36] J. Jayraman and R. de Lima Rodrigues, *J. Phys. A. Math. Gen.*, **23**, 3123, 1990.

[37] M.A. Vasiliev, *Int. J. Mod. Phys.*, **A6**, 1115, 1991.

[38] L. Brink, T.H. Hansson, and M.A. Vasiliev, *Phys. Lett.*, **B286**, 109, 1992.

[39] L. Brink, T.H. Hansson, S. Konstein, and M.A. Vasilev, *Nucl. Phys.*, **B401**, 591, 1993.

[40] T. Brzeziński, I.L. Egusquiza, and A.J. Macfarlane, *Phys. Lett.*, **B311**, 202, 1993.

[41] B. Bagchi, *Phys. Lett.*, **A189**, 439, 1994.

[42] A. Jevicki and J.P. Rodrigues, *Phys. Lett.*, **B146**, 55, 1984.

[43] L.O' Raifeartaigh and C. Ryan, *Proc. Roy. Irish Acad.*, **62**, 93, 1963.

[44] L.M. Yang, *Phys. Rev.*, **84**, 788, 1951.

[45] Y. Ohnuki and S. Kamefuchi, *J. Math. Phys.*, **19**, 67, 1978.

[46] S. Watanabe, *Prog Theor. Phys.*, **80**, 947, 1988.

[47] B. Mielnik, *J. Math. Phys.*, **25**, 3387, 1984.

[48] A. Mitra, A. Lahiri, P.K. Roy, and B. Bagchi, *Int. J. Theor. Phys.*, **28**, 911, 1989.

[49] D.J. Fernandez, *Lett. Math. Phys.*, **8**, 337, 1984.

[50] P.G. Leach, *Physica*, **D17**, 331, 1985.

[51] V. Bargmann, *Rev. Mod. Phys.*, **21**, 488, 1949.

[52] C.V. Sukumar, *J. Phys. A. Math. Gen.*, **18**, 2937, 1985.

[53] D. Baye, *Phys. Rev. Lett.*, **58**, 2738, 1987.

[54] D. Baye and J-M Sparenberg, *Phys. Rev. Lett.*, **73**, 2789, 1994.

[55] D. Baye, *Phys. Rev.*, **A48**, 2040, 1993.

[56] L.U. Ancarani and D. Baye, *Phys. Rev.*, **A46**, 206, 1992.

[57] J-M Sparenberg, *Supersymmetric transformations and the inverse problem in quantum mechanics*, Ph.D. Thesis, University of Brussels, Brussels, 1999.

[58] N. Levinson, *Phys. Rev.*, **75**, 1445, 1949.

[59] R.D. Amado, *Phys. Rev.*, **A37**, 2277, 1988.

[60] G. Levai, D. Baye, and J-M Sparenberg, *J. Phys. A. Math. Gen.*, **30**, 8257, 1997.

[61] B. Talukdar, U. Das, C. Bhattacharyya, and K. Bera, *J. Phys. A. Math. Gen.*, **25**, 4073, 1992.

[62] N. Nag, and R. Roychoudhury, *J. Phys. A. Math. Gen.*, **28**, 3525, 1995.

[63] R.D. Amado, F. Cannata, and J.P. Dedonder, *Int. J. Mod. Phys.*, **A5**, 3401, 1990.

[64] F. Cannata and M.V. Ioffe, *Phys. Lett.*, **B278**, 399, 1992.

[65] C. Eckart, *Phys. Rev.*, **35**, 1303, 1930.

[66] B. Bagchi and R. Roychoudhury, *Mod. Phys. Lett.*, **A12**, 65, 1997.

[67] Q.K.K. Liu, *Nucl. Phys.*, **A550**, 263, 1992.

[68] L.F. Urrutia and E. Hernandez, *Phys. Rev. Lett.*, **51**, 755, 1983.

[69] V.A. Kostelecky and M.M. Nieto, *Phys. Rev.*, **A32**, 1293, 1985.

[70] G. Levai, *Lecture Notes in Physics*, Springer, Berlin, **427**, 127, 1993.

[71] A. Bhattacharjie and E.C.G. Sudarshan, *Nuovo Gim*, **25**, 864, 1962.

[72] A.K. Bose, *Nuovo. Cim.*, **32**, 679, 1964.

[73] W. Miller Jr., *Lie Theory of Special Functions*, Academic Press, New York, 1968.

[74] G.A. Natanzon, *Theor. Math. Phys.*, **38**, 146, 1979.

[75] J.N. Ginocchio, *Ann. Phys.*, **152**, 203, 1984.

[76] J.W. Dabrowska, A. Khare, and U. Sukhatme, *J. Phys. A. Math. Gen.*, **21**, L195, 1988.

[77] A.V. Turbiner, *Commun. Math. Phys.*, **118**, 467, 1988.

[78] M.A. Shifman, *Int. J. Mod. Phys.*, **A4**, 3305, 1989.

[79] O.B. Zaslavskii, *J. Phys. A. Math. Gen.*, **26**, 6563, 1993.

[80] G. Levai, *J. Phys. A. Math. Gen.*, **22**, 689, 1989.

[81] B.W. Williams and D.P. Poulios, *Eur. J. Phys.*, **14**, 222, 1993.

[82] P. Roy, B. Roy, and R. Roychoudhuy, *Phys. Lett.*, **A144**, 55, 1990.

[83] J.M. Cervaro, *Phys. Lett.*, **A153**, 1, 1991.

[84] B.W. Williams, *J. Phys. A. Math. Gen.*, **24**, L667, 1991.

[85] A. Arai, *J. Math. Anal. Appl.*, **158**, 63, 1991.

[86] X.C. Cao, *J. Phys. A. Math. Gen.*, **24**, L1165, 1991.

[87] C.A. Singh and T.H. Devi, *Phys. Lett.*, **A171**, 249, 1992.

[88] G. Levai, *J. Phys. A. Math. Gen.*, **25**, L521, 1992.

[89] P. Cordero, S. Hojman, P. Furlan, and G.C. Ghirardi, *Nuovo Cim*, **A3**, 807, 1971.

[90] P. Cordero and G.C. Ghirardi, *Fortschr Phys.*, **20**, 105, 1972.

[91] G.C. Ghirardi, *Nuovo. Cim*, **A10**, 97, 1972.

[92] G.C. Ghirardi, *Fortschr Phys.*, **21**, 653, 1973.

[93] A. Dey, *A Study of Solvable Potentials in Quantum Mechanics*, dissertation, University of Calcutta, Calcutta, 1993.

[94] B. Bagchi and C. Quesne, *Phys. Lett.*, **A230**, 1, 1997.

[95] A. De Souza Dutra and H. Boschi - Filho, *Phys. Rev.*, **A50**, 2915, 1994.

[96] A. de Souza Dutra, *Phys. Rev.*, **A47**, R2435, 1993.

[97] R. Dutt, A. Khare, and Y.P. Varshni, *J. Phys. A. Math. Gen.*, **28**, L107, 1995.

[98] N. Nag, R. Roychoudhury, and Y.P. Varshni, *Phys. Rev.*, **A49**, 5098, 1994.

[99] C. Grosche, *J. Phys. A. Math. Gen.*, **28**, 5889, 1995.

[100] C. Grosche, *J. Phys. A. Math. Gen.*, **29**, 365, 1996.

[101] G. Junker and P. Roy, *Ann. Phys.*, **270**, 155, 1998.

CHAPTER 6

Radial Problems and Spin-orbit Coupling

6.1 SUSY and the Radial Problems

The techniques of SUSY can also be applied to three and higher dimensional quantum mechanical systems. Consider the three-dimensional problem first. The time-independent Schroedinger equation having a spherically symmetric potential $V(r)$ can be expressed in spherical polar coordinates (r, θ, ϕ) as $(\hbar = m = 1)$

$$
-\frac{1}{2}\left[\frac{1}{r^2}\frac{\partial}{\partial r}\left(r^2\frac{\partial}{\partial r}\right)\right.
$$
$$
+\frac{1}{r^2\sin\theta}\frac{\partial}{\partial\theta}\left(\sin\theta\frac{\partial}{\partial\theta}\right)
$$
$$
\left.+\frac{1}{r^2\sin^2\theta}\frac{\partial^2}{\partial\phi^2}\right]u(r,\theta,\phi)
$$
$$
+V(r)u(r,\theta,\phi) = Eu(r,\theta,\phi) \tag{6.1}
$$

Equation (6.1) is well known [1] to separate in terms of the respective functions of the variables r, θ, and ϕ by writing the wave function as $u(r,\theta,\phi) \equiv R(r)\Theta(\theta)\Phi(\phi)$ where $R(r)$ is the radial part and the angular part $\Theta(\theta)\Phi(\phi)$ is described by the spherical harmonics $Y(\theta,\phi)$.

The radial equation has the form

$$-\frac{1}{2}\frac{1}{r^2}\frac{d}{dr}\left(r^2\frac{dR}{dr}\right) + \left[V(r) - E + \frac{l(l+1)}{2r^2}\right]R = 0 \qquad (6.2)$$

in which the first-order derivative term can be removed by making a further transformation $R \rightarrow \chi(r)/r$. As a result (6.2) can be reduced to

$$-\frac{1}{2}\frac{d^2\chi}{dr^2} + \left[V(r) - E + \frac{l(l+1)}{2r^2}\right]\chi = 0 \qquad (6.3)$$

which is in the form of a Schroedinger equation similar to that of the one-dimensional problems and can be subjected to a supersymmetric treatment. However, for the radial equation $r \in (0, \infty)$ it constitutes only a half-line problem.

To look into how SUSY works in higher-dimensional models let us start with the specific case of the Coulomb potential. We distinguish between two possibilities according to n, the principal quantum number, which is fixed, and l, the angular momentum quantum number, which is allowed to vary [2-4] or as n varies but l is kept fixed [5-7].

Case 1   Fixed n but a variable l

Here $V(r) = -\frac{Ze^2}{r}$. So the radial equation is

$$\left[-\frac{1}{2}\frac{d^2}{dr^2} - E_n - \frac{1}{r} + \frac{l(l+1)}{2r^2}\right]\chi_{nl}(r) = 0 \qquad (6.4)$$

where $\chi_{nl}(0) = 0$, $E_n = -\frac{1}{2n^2}$ and r has been scaled[1] appropriately.

From (2.29) we find that the underlying superpotential for (6.4) satisfies

$$\begin{aligned}V_+ &= \frac{1}{2}\left(W^2 - W'\right) \\ &= -\frac{1}{r} + \frac{l(l+1)}{2r^2} + \frac{1}{2(l+1)^2}\end{aligned} \qquad (6.5)$$

The solution of (6.5) can be worked out as

$$W(r) = \frac{1}{l+1} - \frac{l+1}{r} \qquad (6.6)$$

---

[1]The transformation used is of the form $r \rightarrow \frac{\hbar^2}{\mu}\frac{1}{Ze^2}r$.

implying that the supersymmetric partner to V_+ is

$$
\begin{aligned}
V_- &\equiv \frac{1}{2}\left(W^2 + W'\right) \\
&= -\frac{1}{r} + \frac{(l+1)(l+2)}{2r^2} + \frac{1}{2(l+1)^2}
\end{aligned} \tag{6.7}
$$

Whereas the Bohr series for (6.5) starts from $(l+1)$ with energies $\frac{1}{2}\left[(l+1)^2 - n^{-2}\right]$, we see from (6.7) that the lowest level for V_- begins at $n = l+2$ with $n \geq l+2$. SUSY therefore offers a plausible interpretation of the spectrum of H_+ and H_- which corresponds to the well known hydrogenic $ns - np$ degeneracy. In particular if we set $l = 0$ we find H_+ to describe the ns levels with $n \geq 1$ while H_- is is consistent with the np levels with $n \geq 2$. The main point to note [2-4] is that SUSY brings out a connection between states of same n but different l.

What happens if we similarly apply SUSY to the isotropic oscillator potential $V(r) = \frac{1}{2}r^2$? We find then the Schroedinger equation to read

$$
\left[-\frac{1}{2}\frac{d^2}{dr^2} + \frac{1}{2}r^2 + \frac{l(l+1)}{2r^2} - \left(n + \frac{3}{2}\right)\right] \chi_{nl}(r) = 0 \tag{6.8}
$$

where n is related to l by

$$
n = l, \ l+2, \ l+4, \ldots \tag{6.9}
$$

From (6.8) the superpotential $W(r)$ and the associated supersymmetric partner potentials can be ascertained to be

$$
\begin{aligned}
W(r) &= r - \frac{l+1}{r} \\
V_+(r) &= \frac{1}{2}r^2 + \frac{l(l+1)}{2r^2} - \left(l + \frac{3}{2}\right) \\
V_-(r) &= \frac{1}{2}r^2 + \frac{(l+1)(l+2)}{2r^2} - \left(l + \frac{1}{2}\right)
\end{aligned} \tag{6.10a, b, c}
$$

It is here that we run into a difficulty. The undersirable feature of (6.10b) and (6.10c) is that these only furnish a connection between levels l and $(l+1)$ which is not borne out by (6.9). This counter-example serves as a pointer that SUSY cannot be applied naively to higher-dimensional systems.

It has been argued that supersymmetric transformations are applicable to the radial problems only after the latter have been transformed to the full-line $(-\infty, \infty)$. In the following we see that such an exercise brings out an entirely different role of SUSY; it is found [5,6] to relate states of the same l but different n and nuclear charge Z. Interestingly it does also explain the energy jump of two units in the isotropic oscillator system.

Case 2   Fixed l but a variable n

A relevant transformation which switches $r \in (0, \infty)$ to $x \in (-\infty, \infty)$ is given by $r = e^x$. As a result an equation of the type (6.3) gets transformmed to

$$
-\frac{1}{2}\frac{d^2\Psi}{dx^2} + \left[\{V(e^x) - E\}e^{2x}\right.
$$
$$
\left. + \frac{1}{2}\left(l+\frac{1}{2}\right)^2\right]\Psi = 0 \tag{6.11}
$$

where $\chi(r) \to e^{x/2}\Psi(x)$. Note that E no longer plays the role of an eigenvalue in (6.11).

Corresponding to the Coulomb potential, (6.11) can be written as

$$
-\frac{1}{2}\frac{d^2\Psi}{dx^2} + \left[-e^x - E_n e^{2x}\right.
$$
$$
\left. + \frac{1}{2}\left(l+\frac{1}{2}\right)^2\right]\Psi = 0 \tag{6.12}
$$

which describes a full-line problem for the Morse potential. From (6.12) the superpotential $W(x)$ and $V_\pm(x)$ follow straightforwardly

$$
W(x) = \frac{e^x}{n} + \left(\frac{1}{2} - n\right)
$$
$$
V_+(x) = \frac{e^{2x}}{2n^2} - e^x + \frac{1}{2}\left(\frac{1}{2} - n\right)^2
$$
$$
V_-(x) = \frac{e^{2x}}{2n^2} - \left(1 - \frac{1}{n}\right)e^x + \frac{1}{2}\left(\frac{1}{2} - n\right)^2 \tag{6.13a, b, c}
$$

Having obtained the x-dependent expression for V_- we can now transform back to our old variable r to get the SUSY partner equation. We find in this way [7]

$$
\left[-\frac{1}{2}\frac{d^2}{dr^2} - E_n - \left(1 - \frac{1}{n}\right)\frac{1}{r} \right.
$$
$$
\left. + \frac{l(l+1)}{2r^2} \right] \chi'_{nl}(r) = 0 \tag{6.14}
$$

The nontrivial nature of the mappings $r \to e^x \to r$ is evident from the result that the coefficient of $\frac{1}{r}$ term in (6.14) has undergone a modification by an n-dependent factor compared to (6.4). To interpret (6.14) we therefore need to redefine $\left(1 - \frac{1}{n}\right) r$ as the new variable by dividing (6.14) throughout by the factor of $\left(1 - \frac{1}{n}\right)^2$. This also necessitates redefining the nuclear charge Z by bringing it out explicitly in (6.14): $Z \to Z\left(1 - \frac{1}{n}\right)$. We thus find a degeneracy to hold between states of same l but different n and Z. More specifically, while (6.4) is concerned with states possessing quantum numbers n, l, and energies $-\frac{Z^2}{n^2}\frac{me^4}{\hbar^2}$, (6.14) acconts for states with quantum numbers $(n-1), l$ and nuclear charge $Z(1 - \frac{1}{n})$ having same energies. In this way the model may be used to establish supersymmetric interatomic connections between states of iso-electronic ions under the simultaneous change of the principal quantum number and nuclear charge.

We now consider the isotropic oscillator problem as an application of this scheme.

Here the Schroedinger equation is given by (6.8) along with the energy levels (6.9). By following the prescription of transforming the half-line $(0, \infty)$ to $(-\infty, \infty)$ we employ $x = 2 \ln r$ to get

$$
-\frac{1}{2}\frac{d^2\Psi}{dx^2} + \left[\frac{1}{8}e^{2x} - \frac{1}{4}\left(n + \frac{3}{2}\right)e^x \right.
$$
$$
\left. + \frac{1}{8}\left(l + \frac{1}{2}\right)^2 \right]\Psi = 0 \tag{6.15}
$$

Equation (6.15) gives[2]

$$W(x) = \frac{e^x}{2} - \frac{1}{2}\left(n + \frac{1}{2}\right)$$

$$V_+(x) = \frac{1}{8}e^{2x} - \frac{1}{4}\left(n + \frac{3}{2}\right)e^x + \frac{1}{8}\left(n + \frac{1}{2}\right)^2$$

$$V_-(x) = \frac{1}{8}e^{2x} - \frac{1}{4}\left(n - \frac{1}{2}\right)e^x + \frac{1}{8}\left(n + \frac{1}{2}\right)^2 \qquad (6.16a, b, c)$$

As required we need to transform the Schroedinger equation for V_- back to the half-line to find the suitable supersymmetric partner to (6.8). We find

$$\left[-\frac{1}{2}\frac{d^2}{dr^2} + \frac{1}{2}r^2 + \frac{l(l+1)}{2r^2}\right.$$

$$\left. - \left(n + \frac{3}{2}\right) + 2\right]\chi'_{nl}(r) = 0 \qquad (6.17)$$

It is easy to see that because of the presence of an additional factor of 2 in (6.17) the difference in the energy levels between (6.8) and (6.17) is consistent with (6.9).

6.2 Radial Problems Using Ladder Operator Techniques in SUSYQM

Radial problems can also be handled using ladder operator techniques [8-10]. The advantage is that explicit forms of the superpotential are not necessary. Let us introduce for convenience the ket notation to express the radial equation (6.3) in the form

$$H_l|N,l> \equiv \left[-\frac{1}{2}\frac{d^2}{dr^2} + V(r) + \frac{l(l+1)}{2r^2}\right]|N,l>$$

$$= E_l^N|N,l> \qquad (6.18)$$

Here N denotes the radial quantum number for the Coulomb problem $n = N + l + 1$ while for the isotropic oscillator case $n = 2N + l$, $N = 0, 1, 2, \ldots$ etc.

---

[2]The corresponding eigenvalue of V_+ is $\frac{1}{8}\left[(n + \frac{1}{2})^2 - (l + \frac{1}{2})^2\right]$ so that for $n \geq l$ (and l fixed) the lowest level is zero.

Consider an operator A given by

$$A = \sum_N \alpha_l^N |N', l' >< N, l| \tag{6.19}$$

which obviously maps the ket $|N, l >$ to $|N', l' >$. Then

$$A^+ = \sum_N \beta_l^N |N, l >< N', l'| \tag{6.20}$$

where $\alpha_l^N = (\beta_l^N)^*$.

The representations (6.18) and (6.19) allow one to interpret A^+ and A as raising and lowering operators, respectively

$$
\begin{aligned}
A|N, l > &= \alpha_l^N |N - i, l + j > &\qquad (6.21)\\
A^+|N - i, l + j > &= \beta_l^N |N, l > &\qquad (6.22)
\end{aligned}
$$

where we have set $N' = N - i$ and $l' = l + j$. Furthermore it follows that

$$A^+ A|N, l >= |\alpha_l^N|^2 |N, l >$$

$$AA^+ |N - i, l + j >= |\alpha_l^N|^2 |N - i, l + j > \tag{6.23a, b}$$

where $i, j = 0, 1, 2$ etc.

The next step is to carry out a factorization of the Hamiltonian according to

$$A^+ A = H_l + F$$

$$AA^+ = H_{l+j} + G \tag{6.24a, b}$$

where F and G are scalars independent of the quantum number N. Combining (6.18) with (6.24) we arrive at the result

$$|\alpha_l^N|^2 = E_l^N + F \tag{6.25}$$

Moreover

$$
\begin{aligned}
(H_{l+j} + G) A|N, l > &= AA^+ A|N, l >\\
&= A(E_l^N + F)|N, l > &\qquad (6.26)
\end{aligned}
$$

which implies that

$$H_{l+j} [A|N, l >] = \left\{ E_l^N + F - G \right\} [A|N, l >] \tag{6.27}$$

(6.27) shows $A|N, l>$ to be an eigenket of H_{l+j} as well. But since the eigenvalues of $H_{l'}$ are already known to be $E_{l'}^{N'}$ from (6.18), we have for $N' = N - i$ and $l' = l + j$ the relation

$$E_{l+j}^{N-i} = E_l^N + F - G \tag{6.28}$$

Now by repetitive applications of the operator A on $|N, l>$ a sequence of eigenkets can be created in a manner

$$A^k |N, l> = \alpha_l^N \alpha_{l+j}^{N-i} \ldots \alpha_{l+(k-1)j}^{N-(k-1)i} |N - ki, l + kj> \tag{6.29}$$

where k is a positive integer and indicates how many times A is applied to $|N, l>$. Since this cannot go indefinitely, N being finite, the sequence has to terminate. So we set $A|0, l> = 0$ for $N = 0$ which amounts to restricting $\alpha_l^0 = 0$. The latter fixes $F = -E_l^0$ from (6.25). It is also consistent to choose $i = 1$ since k is a positive integer ≥ 1 and N can take values $0, 1, 2$, etc. As such from (6.28) we have

$$G = \left(E_l^N - E_{l+j}^{N-1} \right) - E_l^0 \tag{6.30}$$

So we are led to a scheme [9,10] of SUSY in which the operators $A^+ A$ and $A A^+$ yield the same spectrum of eigenvalues

$$|\alpha_l^N|^2 = E_l^N - E_l^0 \tag{6.31}$$

corresponding to the eigenkets $|N, l>$ and $|N - i, l + j>$ except for the ground state which satisfies

$$H_l |0, l> = A^+ A |0, l> = 0 \tag{6.32}$$

We can thus define a supersymmetric Hamiltonian analogous to (2.36)

$$H_s = \frac{1}{2} \text{ diag } (A A^+, A^+ A) \equiv \text{ diag } (H_-, H_+) \tag{6.33}$$

with the arguments of (2.55) holding well. In other words, the ground state is nondegenerate (SUSY exact) and is associated with H_+ only. The $H_\pm$ in (6.33) have the representations.

$$H_+ = H_l - E_l^0, \quad H_- = H_{l+j} - G \tag{6.34a, b}$$

with G given by (6.30).

We now turn to some applications of the results (6.34).

(a) Coulomb problem

Here $V(r)$ and E_l^N are

$$V(r) = -\frac{1}{r} \tag{6.35}$$

$$E_l^N = -\frac{1}{2(N+l+1)^2} \tag{6.36}$$

From (6.30) we determine G to be

$$
\begin{aligned}
G =\ & -\frac{1}{2(N+l+1)^2} \\
& +\frac{1}{2(N-1+l+j+1)^2} \\
& +\frac{1}{2(l+1)^2}
\end{aligned}
\tag{6.37}
$$

For G to be N independent it is necessary that $j = 1$. The supersymmetric partner Hamiltonians then read

$$
\begin{aligned}
H_+ =\ & -\frac{1}{2}\frac{d^2}{dr^2} \\
& +\frac{l(l+1)}{2r^2} - \frac{1}{r} \\
& +\frac{1}{2(l+1)^2}
\end{aligned}
\tag{6.38}
$$

$$
\begin{aligned}
H_- =\ & -\frac{1}{2}\frac{d^2}{dr^2} \\
& +\frac{(l+1)(l+2)}{2r^2} - \frac{1}{r} \\
& +\frac{1}{2(l+1)^2}
\end{aligned}
\tag{6.39}
$$

The partner potentials corresponding to the above Hamiltonians are identical to those in (6.5) and (6.7), respectively, and so similar conclusions hold.

(b) Isotropic oscillator problem

In this case the potential and energy levels are given by

$$V(r) = \frac{1}{2}r^2 \tag{6.40}$$

$$E_l^N = 2N + l + \frac{3}{2} \tag{6.41}$$

From (6.30) we find G to be

$$G = 2 - \left(l + j + \frac{3}{2}\right) \tag{6.42}$$

which being free of N does not impose any restriction upon the parameter j. The supersymmetric partner Hamiltonians can be derived from (6.34) to be

$$
\begin{aligned}
H_+ = & -\frac{1}{2}\frac{d^2}{dr^2} \\
& + \frac{l(l+1)}{2r^2} \\
& + \frac{r^2}{2} - \left(l + \frac{3}{2}\right)
\end{aligned} \tag{6.43}
$$

$$
\begin{aligned}
H_- = & -\frac{1}{2}\frac{d^2}{dr^2} \\
& + \frac{(l+j)(l+j+1)}{2r^2} \\
& + \frac{r^2}{2} - \left(l + j + \frac{3}{2}\right) + 2
\end{aligned} \tag{6.44}
$$

A particular case of (6.43) and (6.44) when $j = 0$ corresponds to the previous combinations (6.8) and (6.17). Equations (6.43) and (6.44) are however more general and correctly predicting the energy difference of two units.

Thus using the ladder operator techniques one can successfully give a supersymmetric interpretation to both the Coulomb and the isotropic oscillator problems. We should stress that in getting these results we did not require any specific form of the superpotential. The mere existence of the operators A and A^+ was enough to set up a supersymmetric connection.

6.3 Isotropic Oscillator and Spin-orbit Coupling

Dynamical SUSY can be identified [11,13] for the isotropic oscillator problems involving a constant spin-orbit coupling or when influenced by a particular spin-orbit coupling and an additional potential. Here we discuss the case of spin-orbit coupling only [12].

The underlying Hamiltonian may be taken as

$$H = \frac{1}{2}\left(-\frac{d^2}{dr^2} + r^2\right) + \lambda\left(\vec{\sigma}.\vec{L} + \frac{3}{2}\right) \tag{6.45}$$

having energy levels

$$
\begin{aligned}
E^\lambda_{nl_j} &= 2N + l + \frac{3}{2} + \lambda\left(l + \frac{3}{2}\right) \text{ for } j = l + \frac{1}{2} \\
&= 2N + l + \frac{3}{2} + \lambda\left(-l + \frac{1}{2}\right) \text{ for } j = l - \frac{1}{2}
\end{aligned} \tag{6.46}
$$

where N as usual denotes the radial quantum number $(N = 0, 1, 2, \ldots)$ for a fixed j and λ is a coupling constant. Equation (6.46) reflects an obvious degeneracy for $\lambda = -1$, $l = j + \frac{1}{2}$ and $\lambda = 1$, $l = j - \frac{1}{2}$. In both cases E_N turns out to be

$$E_N = 2N + 2j + 2 \tag{6.47}$$

To show that a dynamical SUSY is associated with the Hamiltonian we consider first an $SU(1,1)$ algebra in terms of the operators $K_\pm$ and K_0

$$
\begin{aligned}
{[K_\pm, K_0]} &= \mp K_\pm \\
{[K_+, K_-]} &= -2K_0
\end{aligned} \tag{6.48}
$$

A convenient set of representations may be adopted for (6.48) which is

$$K_+ = \frac{1}{2}\sum_{i=1}^{3}\beta_i^+\beta_i^+$$

$$K_- = \frac{1}{2}\sum_{i=1}^{3}\beta_i\beta_i$$

$$K_0 = \frac{1}{2}\sum_{i=1}^{3}\left(\beta_i^+\beta_i + \frac{1}{2}\right) \tag{6.49}$$

where β_i and β_i^+ satisfy the bosonic commutation relations

$$\left[\beta_i, \beta_j^+\right] = \delta_{ij}, \quad i, j = 1, 2, 3 \tag{6.50}$$

Actually one can even enlarge the algebra (6.48) by introducing a set of operators F_+ and F_- which are

$$F_+ = \frac{1}{2}\sum_{i=1}^{3}\sigma_i\beta_i^+ \begin{pmatrix} 0 & 1 \\ 1 & 0 \end{pmatrix}$$

$$F_- = \frac{1}{2}\sum_{i=1}^{3}\sigma_i\beta_i \begin{pmatrix} 0 & 1 \\ 1 & 0 \end{pmatrix} \tag{6.51}$$

Then the following commutation relations enlarge

$$\begin{aligned} \left[F_\pm, K_0\right] &= \mp\frac{1}{2}F_\pm \\ \left[K_\pm, F_\mp\right] &= \mp F_\pm \\ \left[K_+, F_+\right] &= 0 \\ \left[K_-, F_-\right] &= 0 \end{aligned} \tag{6.52}$$

along with

$$\begin{aligned} \{F_\pm, F_\pm\} &= K_\pm \\ \{F_+, F_-\} &= K_0 \end{aligned} \tag{6.53}$$

Equations (6.48), (6.52), and (6.53) constitute the noncompact $Osp(\frac{1}{2})$ superalgebra.

The quadratic Casimir operator of Osp $(\frac{1}{2})$ and $SU(1,1)$ can be defined in terms of the commutation of F_+ and F_- which we call $\widehat{C}$. Thus

$$C_2\left[Osp(\frac{1}{2}\right] = \widehat{C}^2 + \frac{1}{2}\widehat{C} \tag{6.54}$$

$$C_2\left[SU(1,1)\right] = \widehat{C}^2 + \widehat{C} \tag{6.55}$$

where

$$\widehat{C} = [F_+, F_-] \tag{6.56}$$

Now corresponding to (6.51) $\widehat{C}$ can be found to be

$$\widehat{C} = \frac{1}{4}\sum_i\sum_j\left[\beta_i^+\beta_j\left(\sigma_i\sigma_j - \sigma_j\sigma_i\right)\right] - 3 \tag{6.57}$$

which implies

$$\widehat{C} = -\frac{1}{2}\left(\vec{\sigma}.\vec{L}\right) - \frac{3}{4} \tag{6.58}$$

where note that $\vec{L} = i(\beta_2\beta_3^+ - \beta_2^+\beta_3, \text{ cyclic})$.

Using (6.58), the Casimirs for Osp $\left(\frac{1}{2}\right)$ and $SU(1,1)$ become

$$C_2\left[Osp\left(\frac{1}{2}\right)\right] = \frac{1}{4}\left(\vec{L} + \frac{1}{2}\vec{\sigma}\right)^2 \tag{6.59}$$

$$C_2\left[SU(1,1)\right] = \frac{1}{2}\vec{L}^2 - \frac{3}{16} \tag{6.60}$$

while the Hamiltonian (6.45) takes the form

$$H = 2K_0 - 2\lambda\widehat{C} \tag{6.61}$$

In fact H can be written in terms of the Casimir operators of Osp $\left(\frac{1}{2}\right)$ and $SU(1,1)$ by inserting (6.59) and (6.60) in (6.61).

$$H = 4\lambda\left[C_2\left(Osp\left(\frac{1}{2}\right)\right) - C_2\left(SU(1,1)\right)\right] + 2K_0 \tag{6.62}$$

Equation (6.62) reflects a dynamical SUSY corresponding to the supergroup embedding Osp $\left(\frac{1}{2}\right) \supset SU(1,1) \supset SO(2)$.

To explicitly bring out the connection of the Hamiltonian (6.45) to SUSYQM one has to take recourse to defining some additional operators. These are the U's, W's, and Y which along with (6.48), (6.52) and (6.53) enlarge Osp $\left(\frac{1}{2}\right)$ to the Osp $\left(\frac{2}{2}\right)$ superalgebra. Their representations can be taken to be

$$U_+ = \frac{1}{\sqrt{2}}\begin{pmatrix} 0 & \sigma_i\beta_i^+ \\ 0 & 0 \end{pmatrix}$$

$$U_- = \frac{1}{\sqrt{2}}\begin{pmatrix} 0 & \sigma_i\beta_i \\ 0 & 0 \end{pmatrix}$$

$$W_+ = \frac{1}{\sqrt{2}}\begin{pmatrix} 0 & 0 \\ \sigma_i\beta_i^+ & 0 \end{pmatrix}$$

$$W_- = \frac{1}{\sqrt{2}}\begin{pmatrix} 0 & 0 \\ \sigma_i\beta_i & 0 \end{pmatrix}$$

$$Y = \frac{1}{2}\left(\frac{3}{2} + \vec{\sigma}.\vec{L}\right)\begin{pmatrix} 1 & 0 \\ 0 & -1 \end{pmatrix} \tag{6.63}$$

where a summation over the label i is suggested. One then finds the following relations to hold

$$
\begin{aligned}
[U_\pm, K_0] &= \mp\frac{1}{2}U_\pm \\
[W_\pm, K_0] &= \mp\frac{1}{2}W_\pm \\
[K_\pm, U_\pm] &= 0, \\
[K_\pm, W_\pm] &= 0 \\
[K_\pm, U_\mp] &= \mp U_\pm \\
[K_\pm, W_\mp] &= \mp W_\pm \\
[Y, U_\pm] &= \frac{1}{2}U_\pm \\
[Y, W_\pm] &= -\frac{1}{2}W_\pm \\
[K_\pm, Y] &= 0 \\
[K_0, Y] &= 0 \\
\{U_\pm, U_\pm\} &= 0 \\
\{W_\pm, W_\pm\} &= 0 \\
\{U_+, U_-\} &= 0 \\
\{W_+, W_-\} &= 0 \\
\{U_\pm, W_\pm\} &= K_\pm \\
\{U_\mp, W_\pm\} &= K_0 \pm Y
\end{aligned}
\tag{6.64}
$$

Moreover corresponding to the two cases $\lambda = \pm1, l = j \mp \frac{1}{2}$, the Hamiltonians (6.45) are expressible as

$$
H = 2(K_0 + Y) \tag{6.65}
$$

The degneracy indicated by (6.47) can be understood [12] from the fact that both the Hamiltonians belong to the same representation of Osp $\left(\frac{2}{2}\right) \sim SU(1,1/1)$. That (6.65) can be put in the supersymmetric form follows from the identifications $A \sim \sqrt{2}\sum\sigma_i\beta_i^+$ and $A^+ \sim \sqrt{2}\sum\sigma_i\beta_i^+$. In our notations of Chapter 2 these mean $U_- = Q/\sqrt{2}$ and $W_+ = Q^+/\sqrt{2}$, so that from the last equation of (6.64) corresponding to the positive sign we have $H = \{Q, Q^+\}$.

6.4   SUSY in D Dimensions

The radial Schroedinger equation in D dimensions reads (see Appendix A for a detailed derivation)

$$\left[-\frac{1}{2}\frac{d^2}{dr^2} - \frac{D-1}{2r}\frac{d}{dr} + \frac{l(l+D-2)}{2r^2} + V(r)\right]R = ER \qquad (6.66)$$

where r in terms of D cartesian coordinates x_i is given by $r = \left[\sum_{i=1}^{D} x_i^2\right]^{\frac{1}{2}}$. As with (6.2), here, too, the first order derivative term can be removed by employing the transformation $R \to r^{-\frac{D-1}{2}}\chi(r)$. We then have the form

$$\left[-\frac{1}{2}\frac{d^2}{dr^2} + \frac{\alpha_l}{2r^2} + V(r)\right]\chi = E\chi \qquad (6.67)$$

where

$$\alpha_l = \frac{1}{4}(D-1)(D-3) + l(l+D-2) \qquad (6.68)$$

We now consider the following cases

(a)   Coulomb potential

The energy spectrum corresponding to the Coulomb potential $V(r) = -\frac{1}{r}$ is

$$E_l^N = -\frac{1}{2}\frac{1}{\left[N+l+\left(\frac{D-1}{2}\right)\right]^2} \qquad (6.69)$$

In (6.69) N and l stand for the radial and angular momentum quantum numbers respectively.

From (6.30) we obtain

$$\begin{aligned}
G &= -\frac{1}{2}\frac{1}{\left[N+l+\left(\frac{D-1}{2}\right)\right]^2}\\
&\quad + \frac{1}{2}\frac{1}{\left[l+\left(\frac{D-1}{2}\right)\right]^2}\\
&\quad + \frac{1}{2}\frac{1}{\left[N-1+l+j+\left(\frac{D-1}{2}\right)\right]^2} \qquad (6.70)
\end{aligned}$$

Setting $j = 1$ it is easy to realize that G becomes independent of N leading to

$$G = \frac{1}{2} \frac{1}{\left[l + \left(\frac{D-1}{2}\right)\right]^2} \tag{6.71}$$

Then the general results for $H_{\pm}$ in the D dimensional space are

$$
\begin{aligned}
H_+ &\equiv H_l - E_l^0 = -\frac{1}{2}\frac{d^2}{dr^2} + \frac{\alpha_l}{2r^2} \\
&\quad -\frac{1}{r} + \frac{1}{2}\left[l + \frac{1}{2}(D-1)\right]^{-2} \tag{6.72}
\end{aligned}
$$

$$
\begin{aligned}
H_- &\equiv H_{l+1} + G = -\frac{1}{2}\frac{d^2}{dr^2} + \frac{\alpha_{l+1}}{2r^2} \\
&\quad -\frac{1}{r} + \frac{1}{2}\left[l + \frac{1}{2}(D-1)\right]^{-2} \tag{6.73}
\end{aligned}
$$

where α_l is given by (6.68). For $D = 3$ these are in agreement with (6.38) and (6.39).

(b) Isotropic oscillator potential

For the isotropic oscillator potential $V(r) = \frac{1}{2}r^2$ the energy levels are

$$E_l^N = \left(2N + l + \frac{1}{2}D\right), \quad D \geq 2 \tag{6.74}$$

From (6.30) we find

$$G = 2 - \left(l + j + \frac{D}{2}\right) \tag{6.75}$$

which turns out to be independent of N. It gives the following isospectral Hamiltonians

$$
\begin{aligned}
H_+ &= -\frac{1}{2}\frac{d^2}{dr^2} + \frac{\alpha_l}{2r^2} \\
&\quad +\frac{1}{2}r^2 - \left(l + \frac{D}{2}\right) \tag{6.76}
\end{aligned}
$$

$$
\begin{aligned}
H_- &= -\frac{1}{2}\frac{d^2}{dr^2} + \frac{\alpha_{l+j}}{2r^2} \\
&\quad +\frac{1}{2}r^2 - \left(l + j + \frac{D}{2}\right) + 2 \tag{6.77}
\end{aligned}
$$

These may be compared with (6.43) and (6.44) for $D = 3$.

To conclude it is worthwhile to note that transformations from the Coulomb problem to the isotropic oscillator and vice-versa can be carried out [13-23] and the results turn out to be generalizable to D-dimensions [13-24].

6.5 References

[1] B.H. Bransden and C.J. Joachain, *Introduction to Quantum Mechanics, ULBS*, Longman Group, Essex, UK, 1984.

[2] V.A. Kostelecky and M.M. Nieto, *Phys. Rev. Lett.*, **53**, 2285, 1984.

[3] V.A. Kostelecky and M.M. Nieto, *Phys. Rev. Lett.*, **56**, 96, 1986.

[4] V.A. Kostelecky and M.M. Nieto, *Phys. Rev.*, **A32**, 1293, 1985.

[5] R.W. Haymaker and A.R.P. Rau, *Am. J. Phys.*, **54**, 928, 1986.

[6] A.R.P. Rau, *Phys. Rev. Lett.*, **56**, 95, 1986.

[7] A. Lahiri, P.K. Roy, and B. Bagchi, *J. Phys. A. Math. Gen.*, **20**, 3825, 1987.

[8] J.D. Newmarch and R.H. Golding, *Am. J. Phys.*, **46**, 658, 1978.

[9] A. Lahiri, P.K. Roy, and B. Bagchi, *Phys. Rev.*, **A38**, 6419, 1988.

[10] A. Lahiri, P.K. Roy, and B. Bagchi, *Int. J. Theor. Phys.*, **28**, 183, 1989.

[11] A.B. Balantekin, *Ann. Phys.*, **164**, 277, 1985.

[12] H. Ui and G. Takeda, *Prog. Theor. Phys.*, **72**, 266, 1984.

[13] V.A. Kostelecky, M.M. Nieto, and D.R. Truax, *Phys. Rev.*, **D32**, 2627, 1985.

[14] B. Baumgartner, H. Grosse, and A. Martin, *Nucl. Phys.*, **B254**, 528, 1985.

[15] A. Lahiri, P.K. Roy, and B. Bagchi, *J. Phys. A: Math. Gen.*, **20**, 5403, 1987.

[16] J.D. Louck and W.H. Schaffer, *J. Mol. Spectrosc*, **4**, 285, 1960.

[17] J.D. Louck, *J. Mol. Spectrosc*, **4**, 298, 1960.

[18] J.D. Louck, *J. Mol. Spectrosc*, **4**, 334, 1960.

[19] D. Bergmann and Y. Frishman, *J. Math. Phys.*, **6**, 1855, 1965.

[20] D.S. Bateman, C. Boyd, and B. Dutta-Roy, *Am. J. Phys.*, **60**, 833, 1992.

[21] P. Pradhan, *Am. J. Phys.*, **63**, 664, 1995.

[22] H.A. Mavromatis, *Am. J. Phys.*, **64**, 1074, 1996.

[23] A. De, *Study of a Class of Potential in Quantum Mechanics*, Dissertation, University of Calcutta, Calcutta, 1997.

[24] A. Chatterjee, *Phys. Rep.*, **186**, 249, 1990.

CHAPTER 7

Supersymmetry in Nonlinear Systems

7.1 The KdV Equation

One of the oldest known evolution equations is the KdV, named after its discoverers Korteweg and de Vries [1], which governs the motion of weakly nonlinear long waves. If $\delta(x, t)$ is the elevation of the water surface above some equilibrium level h and α is a parameter characterizing the motion of the medium, then the dynamics of the flow can be described by an equation of the form

$$\delta_t = \frac{3}{2}\sqrt{\frac{g}{h}}\left(\delta\delta_x + \frac{2}{3}\alpha\delta_x + \frac{1}{3}\sigma\delta_{xxx}\right) \tag{7.1}$$

where the suffixes denote partial derivatives with respect to the space (x) and time (t) variables. The parameter σ signifies the relationship between the surface tension T of the fluid and its density $\rho : \frac{h^3}{3} - \frac{hT}{\rho g}$.

It can be easily seen that (7.1) can be put in a more elegant form

$$u_t = 6uu_x - u_{xxx} \tag{7.2}$$

by a simple transformation of the variable δ and scaling the parameters h, α, and σ appropriately. In the literature (7.2) is customarily referred to as the KdV equation. Some typical features which (7.2) exhibit are

(i)   Galilean invariance: The transformations $u'(x', t') \rightarrow u(x, t) + \frac{u_0}{6}$ where $x' \rightarrow x \pm u_0 t$ and $t' \rightarrow t$ leave the form of (7.2) unchanged.

(ii)   PT symmetry: Both $u(x, t)$ and $u(-x, -t)$ are solutions of (7.2).

(iii)   Solitonic solution: Equation (7.2) possesses a solitary wave solution

$$u(x, t) = -\frac{a}{2} \mathrm{sech}^2 \left[\frac{\sqrt{a}}{2} (x - at) \right], \quad a \in \mathbb{R} \qquad (7.3)$$

which happens to be a solitonic solution as well.

Solitary waves in general occur due to a subtle interplay between the steepening of nonlinear waves and linear dispersive effects [2]. Sometimes solitary waves are also form-preserving, these are then referred to as solitons. Solitons reflect particle-like behaviour in that they proceed almost freely, can collide among themselves very much like the particles do, and maintain their original shapes and velocities even after mutual interactions are over.

It was Scott-Russell who first noted a solitary wave while observing the motion of a boat. In a classic paper [3], Scott-Russell gives a fascinating account of his chance meeting with the solitary wave in the following words

. . . I was observing the motion of a boat which was rapidly drawn along a narrow channel by a pair of horses, when the boat suddenly stopped (but) not so the mass of the water in the channel which it had put in motion; it accumulated around the prow of the vessel in a state of violent agitation, then suddenly leaving it behind, rolled forward with great velocity, assuming the form of a large solitary elevation, a rounded, smooth and well-defined heap of water, which continued its course along the channel apparently without change of form or diminution of speed.

We now know that what Scott-Russell saw was actually a solitary wave. We also believe that the KdV equation provides an analytical basis to his observations.

Intense research on the KdV equation began soon after Gardner and Morikawa [4] found an application in the problem of collision-free hydromagnetic waves. Subsequently works on modelling of longitudinal waves in one-dimensional lattice of equal mass were reported [5,6] and numerical computations were carried out [7] to compare with the recurrences observed in the Fermi-Pasta-Ulam model [8]. The relevance of KdV to describe pressure waves in a liquid gas bubble chamber was also pointed out [9]. Moreover, the KdV was found to play an important role in explaining the motion of the three-dimensional water wave problem [10]. A number of theoretical achievements were made alongside these. Gardner, Greene, Kruskal, and Miura [11,12] developed a method of finding an exact solution for the initial-value problem. Lax [13] proposed an operator approach towards interpreting nonlinear evolution equations in terms of conserved quantities, and Zakharov and Shabat [14] formulated what is called the inverse scattering approach. Further, it was also discovered [15,16] that algebraic connections could be set up by means of Backlund transformations implying that the solutions of certain evolutionary equations are correlated. We do not intend to go into the details of the various research directions which took off from these works but we will focus primarily on the symmetry principles like conservation laws as well as the solutions of a few nonlinear equations which can now be analyzed in terms of the concepts of SUSY. All this will form the subject matter of this chapter.

7.2 Conservation Laws in Nonlinear Systems

The existence of conservation laws leads to integrals of motion. A conservation law is an equation of the type

$$T_t + \chi_x = 0 \qquad (7.4)$$

where T is the conserved density and $-\chi$ represents the flux of the flow. The KdV equation is expressible in the form (7.4), a property it shares with several other evolution equations like the modified KdV (MKdV), Sine-Gordon (SG), Liouville equation, etc. Obviously in all these equations conservation laws exist.

Let us show how conservation laws can be derived for the KdV equation. Following the treatment of Miura, Gardner, and Kruskal

[17] we express $u(x, t)$ in terms of a function $w(x, t)$ defined by

$$u = w + \epsilon w_x + \epsilon^2 w^2 \qquad (7.5)$$

where ϵ is a parameter. The left-hand-side of (7.2) can be factorized in a manner

$$u_t - 6uu_x + u_{xxx} = \left(1 + \epsilon \frac{\partial}{\partial x} + 2\epsilon^2 w\right)$$
$$\left[w_t - 6\left(w + \epsilon^2 w^2\right) w_x + w_{xxx}\right] \qquad (7.6)$$

implying that if u is a solution of the KdV equation then the function w needs to satisfy

$$w_t - 6\left(w + \epsilon^2 w^2\right) w_x + w_{xxx} = 0 \qquad (7.7)$$

(7.7) is called the Gardner equation. We may rewrite (7.7) as

$$w_t + \left(-3w^2 - 2\epsilon^2 w^3 + w_{xx}\right)_x = 0 \qquad (7.8)$$

and look for an expression of w like

$$w = \Sigma_{n=0}^{\infty} \epsilon^n w_n(u) \qquad (7.9)$$

Substitution of (7.9) in (7.5) gives the correspondence

$$\begin{aligned} w_0 &= u \\ w_1 &= -w_{0x} = -u_x \\ w_2 &= -w_{1x} - w_0^2 = u_{xx} - u^2 \end{aligned} \qquad (7.10)$$

etc.

On the other hand, combining (7.8) with (7.9) and matching for the coefficients ϵ, ϵ^2, and so on we are also led to

$$\begin{aligned} w_{0t} + \left(-3w_0^2 + w_{0xx}\right)_x &= 0 \\ w_{1t} + (-6w_0 w_1 + w_{1xx})_x &= 0 \\ w_{2t} + \left(-6w_0 w_2 - 3w_1^2 - 2w_0^3 + w_{2xx}\right)_x &= 0 \end{aligned} \qquad (7.11)$$

etc.

In this way we see that the KdV admits of an infinity of conservation laws. Exploiting (7.10), the conserved densities and flows for the KdV can be arranged as follows

$$T_0 = u$$

$$\chi_0 = u_{xx} - u^2$$

$$T_1 = u^2$$

$$\chi_1 = 2uu_{xx} - 4u^3 - u_x^3$$

$$T_2 = u^3 + \frac{1}{2}u_x^2$$

$$\chi_2 = -\frac{9}{2}u^4 + 3u^2 u_{xx} - 6uu_x^2 + u_x u_{xxx} - \frac{1}{2}u_{xx}^2 \qquad (7.12)$$

etc.

Note that the set (T_0, χ_0) corresponds to the KdV equation itself. However, (T_1, χ_i), $i = 1, 2, \ldots$ yield higher order KdV equations. Indeed a recursion operator can be found [18] by which the infinite hierarchy of the corresponding equations can be derived.

To touch upon other evolution equations it is convenient to reformulate the inverse method in terms of the following coupled system involving the functions $\Psi(x, t)$ $\Phi(x, t)$

$$\Psi_x - \lambda\Psi = f\Phi$$
$$\Phi_x + \lambda\Phi = g\Psi \qquad (7.13)$$

where f and g also depend upon x and t. The question we ask is under what conditions the eigenvalues λ are rendered time-independent.

Suppose that the time evolutions of Ψ and Φ are given by the forms

$$\Psi_t = a(x, t, \lambda)\Psi + b(x, t, \lambda)\Phi$$
$$\Phi_t = c(x, t, \lambda)\Psi - a(x, t, \lambda)\Phi \qquad (7.14a, b)$$

where as indicated a, b, c, d are certain functions of x, t, and λ. The conditions for the time-independence of λ may be worked out to be [19]

$$a_x = fc - gb$$
$$b_x - 2\lambda b = f_t - 2af$$
$$c_x + 2\lambda c = g_t + 2ag \qquad (7.15)$$

By taking different choices of a, b, and c the corresponding evolution equations can be derived. A few are listed in Table 7.1.

The structure of the conservation laws following from (7.13), and (7.14) can be made explicit if we introduce the quantities ξ and η to be

$$\xi = \frac{\Phi}{\Psi}, \; \eta = \frac{\Psi}{\Phi} \tag{7.16}$$

The set of equations (7.13) and (7.14) can be expressed in terms of these variables as

$$\begin{aligned}
\xi_x &= g - 2\lambda\xi - f\xi^2 \\
\xi_t &= c - 2a\xi - b\xi^2 \\
\eta_x &= f + 2\lambda\eta - g\eta^2
\end{aligned}$$

$$\eta_t = b + 2a\eta - c\eta^2 \tag{7.17a, b, c, d}$$

Eliminating c between (7.17b) and (7.14b) we get

$$a + b\xi = \frac{1}{\Psi}\frac{\partial\Psi}{\partial t} \tag{7.18}$$

This implies

$$\frac{\partial}{\partial x}(a + b\xi) = \frac{\partial}{\partial t}(f\xi) \tag{7.19a}$$

Similarly eliminating b between (7.17d) and (7.14a) we arrive at

$$\frac{\partial}{\partial x}(-a + c\eta) = \frac{\partial}{\partial t}(g\eta) \tag{7.19b}$$

(7.19a) and (7.19b) are the required conservation relations.

We have already provided the first few conserved densities and the fluxes for the KdV. We now furnish the same for the MKdV and SG equations. Note that the generalized form of the MKdV equation is

$$v_t + 6(k^2 - v^2)v_x + u_{xxx} = 0 \tag{7.20}$$

where k is a constant. On the other hand, the SG equation when expressed in light-cone coordinates $x_\pm = \frac{1}{2}(x \pm t)$ reads

$$\frac{\partial^2\psi}{\partial x_+ \partial x_-} = -\sin\psi \tag{7.21a}$$

Table 7.1

A summary of nonlinear equations KdV, MKdV and SG for various choices of the coefficient functions defined in the text

Equation	f	g	a	b	c
KdV : $u_t = 6uu_x - u_{xxx}$	$-u$	-1	$-4\lambda_3$	$u_{xx} + 2\lambda u - 2u^2$	$4\lambda^2 - 2u$
MKdV : $v_t = 6v^2v_x - v_{xxx}$	$-v$	v	$-4\lambda^3 - 2\lambda v^2$	$v_{xx} + 2\lambda v_x + 4\lambda^2 v + 2v^3$	$-v_{xx} + 2\lambda v_x - 4\lambda^2 v - 2v^3$
SG : $\psi' = -\sin\psi$	$\frac{1}{2}u_x$	$-\frac{1}{2}u_x$	$\frac{1}{\lambda}\cos u$	$\frac{1}{4\lambda}\sin u$	$\frac{1}{4\lambda}\sin u$

That is

$$\dot{\psi}' = -\sin\psi \qquad (7.21b)$$

where the overdot (prime) denotes a partial derivative with respect to $x_+(x_-)$.

The results for (T_i, χ_i) corresponding to (7.20) and (7.21) are

MKdV

$$
\begin{aligned}
T_1 &= v^2, \\
\chi_1 &= -3v^4 - 2vv_{xx} + v_x^2
\end{aligned}
$$

$$
\begin{aligned}
T_2 &= v^4 - 6k^2v^2 + v_x^2, \\
\chi_2 &= v^6 + v^3 v_{xx} - 3v^2 v_x^2 + \frac{v_x v_{xxx}}{2} - \frac{v_x^2}{4}
\end{aligned}
\qquad (7.22)
$$

etc.

SG

$$
\begin{aligned}
T_0 &= \psi'^2 \\
\chi_0 &= 2\cos\psi
\end{aligned}
$$

$$
\begin{aligned}
T_1 &= \psi'^4 - 4\psi''^2, \\
\chi_1 &= 4\psi'^2 \cos\psi
\end{aligned}
$$

$$
\begin{aligned}
T_2 &= \psi'^6 - 4\psi'^2 \psi''^2 + \frac{16}{3}\psi'^3 \psi''' + 8\psi'''^2 \\
\chi_2 &= 2\left(\frac{1}{3}\psi'^4 - 4\psi''^2\right)\cos\psi
\end{aligned}
\qquad (7.23)
$$

etc.

Now given a conservation law in the form (7.4) we can identify the corresponding constants of motion in the manner

$$I[f] = \int_{-\infty}^{\infty} T\left[f(x,t)\right] dx \qquad (7.24)$$

where $f(x,t)$ and its derivatives are assumed to decrease to zero sufficiently rapidly, that is as $|x| \to \infty$. Here is a summary of $I[f]$ for the KdV, MKdV, and SG systems

KdV

$$I_0 = \int dx\, u$$

$$I_1 = \int dx\, u^2$$

$$I_2 = \int dx \left(u^3 + \frac{1}{2} u_x^2 \right) \tag{7.25}$$

etc.

MKdV

$$I_1 = \int dx\, v^2$$

$$I_2 = \int dx \left(v^4 + v_x^2 - 6k^2 v^2 \right) \tag{7.26}$$

etc.

SG

$$I_0 = \int dx_-\, \psi'^2$$

$$I_1 = \int dx_- \left(\psi'^4 - 4\psi''^2 \right)$$

$$I_2 = \int dx_- \left(\psi'^6 - 20\psi'^2\psi''^2 + 8\psi'''^2 \right) \tag{7.27}$$

etc.

In writing I_2 of SG we have integrated by parts and discarded a total derivative.

7.3 Lax Equations

Lax's idea of an operator formulation [13] of evolution equations gives much insight into the rich symmetry structure of nonlinear systems. The KdV equation admits a Lax representation which implies that (7.2) can be represented as

$$L_t = [L, B] \tag{7.28}$$

where the operators L and B are given by

$$L = -\frac{\partial^2}{\partial x^2} + u(x,t)$$

$$B = 4\frac{\partial^3}{\partial x^3} - 3\left(u\frac{\partial}{\partial x} + \frac{\partial}{\partial x}u\right) \qquad (7.29a,b)$$

Equation (7.28) can be solved to obtain

$$L(t) = S(t)L(0)S^{-1}(t)$$

$$\dot{S} = -BS \qquad (7.30a,b)$$

Corresonding to (7.29a) the associated eigenvalue problem is $L\Phi = \lambda\Phi$ which means that if Φ is an initial eigenfunction of L with an eigenvalue λ then it remains so for all times bearing the same eigenvalue λ. The essence of (7.30a) and $L\Phi = \lambda\Phi$ is that the spectrum of L is conserved and that it yields an infinite sequence of conservation laws. Note that the conserved quantities may also be obtained from the definitions

$$L(x,y) = \left[-\frac{\partial^2}{\partial x^2} + u(x)\right]\delta(x-y) \qquad (7.31a)$$

along with

$$TrL = \int dx dy \delta(x-y)L(x,y) \qquad (7.31b)$$

Now since

$$L^2(x,y) = \int dz L(x,z)L(x,y)$$

$$= \left[\frac{\partial^4}{\partial x^4} - \{u(x) + u(y)\}\frac{\partial^2}{\partial x^2} + u^2\right]\delta(x-y) \qquad (7.32a)$$

we obtain

$$TrL = -v\delta''(0) + \delta(0)\int dx u(x) \qquad (7.32b)$$

$$TrL^2 = -V\delta'''(0) - 2\delta''(0)\int dx u + \delta(0)\int dx u^2 \qquad (7.32c)$$

where $V \equiv \int_{-\infty}^{\infty} dx$. If we agree to ignore the additive and multiplicative infinities, TrL and TrL^2 lead to the conserved quantities

$(7.25a, b)$. The extraction of the conserved quantities when the conserved functional contains more than one term, as with I_2 and I_3 etc., is a little tricky. One invokes the rule of counting each power of u to be equivalent to two derivatives. Then introducing an arbitrary coefficient with each term which are determined from $\dot{I}_i = 0 (i = 2, 3, \ldots)$ the higher conservation laws can be obtained [see Chodos [20,21] who also discusses another method of determining the conservation laws using pseudo-differential operators.]

Like the KdV the SG equation can also be cast in the Lax form (7.28) with

$$L = 2\sigma_3 \frac{\partial}{\partial x_-} + \sigma_2 \psi'$$

$$B = (\sigma_3 \cos \psi + \sigma_2 \sin \psi) L^{-1} \tag{7.33}$$

The use of light-cone coordinates reflects that the conserved quantities will turn out to be the x_- integrals of appropriate functions of ψ.

7.4 SUSY and Conservation Laws in the KdV-MKdV Systems

An interesting aspect of $(7.29a)$ is that its time-independent version furnishes the Schroedinger equation

$$\left[-\frac{d^2}{dx^2} + u(x, 0) \right] \Phi = \lambda \Phi \tag{7.34}$$

having the stationary solution of the KdV equation as the potential. Thus it is through the L operator that one looks for a correspondence between the KdV and the Schroedinger eigenvalue problem. In Chapter 2, we noted that the N soliton solutions of the KdV equation emerge as families of reflectionless potentials [22-25]. We may write them in the form (for $t = 0$) : $u_N(x, t) = -N(N+1) b^2 \text{sech}^2 bx$, $N = 1, 2, \ldots$, which is a family of symmetric, reflectionless potentials. The case $N = 1, b = \frac{\sqrt{a}}{2}$ corresponds to the one-soliton solution for $t = 0$, while the one for all t is given by (7.3).

Turning now to the MKdV equation we notice that the quantities

$$u_\pm = \frac{1}{2}(v^2 \mp v_x - k^2) \tag{7.35}$$

where k is a constant satisfy the KdV equation provided the condition

$$2v \left[v_t + 6(k^2 - v^2)v_x + v_{xxx} \right] \mp \left[v_t + 6(k^2 - v^2)v_x + v_{xxx} \right]_x = 0 \tag{7.36}$$

holds, in other words if v is a solution of the MKdV equation. Actually one of the solutions in (7.35) corresponds (when $k = 0$) to the well-known Miura-map between the KdV and MKdV. Observe that the MKdV equation is invariant under $v \to -v$ so both $u_\pm$ correlate the KdV and MKdV. In this sense the Miura transformation can be viewed as the supersymmetric square-root [26].

The combinations (7.35) remind one of the partner potentials $V_\pm$ encountered in Chapter 2. Here v plays the role of the superpotential. Interestingly, using (7.35) it is possible to work out the conserved quantities for the MKdV from those of the KdV equation. Indeed employing (7.35) it is straightforward to verify that the conserved quantities $I_0, I_1, \ldots$ of the KdV get mapped to $I_1, I_2, \ldots$ of the MKdV.

Physically the transformations (7.35) mean that if we define

$$V_- \equiv \frac{1}{2} \left(v^2 + v_x \right) = \frac{1}{2} k^2 > 0 \tag{7.37}$$

then, as in (2.71), V_- does not have any bound state. On the other hand, V_+ reading $V_+ \equiv \frac{1}{2} \left(v^2 - v_x \right)$ reveals a zero-energy bound state. Thus one can carry out the construction of reflectionless potentials as outlined in Chapter 2. Furthermore, in the spirit of SUSY, employing appropriate boundary conditions on v, the solutions u_+ and u_- may be identified with the $N + 1$ and N soliton solutions, respectively.

7.5 Darboux's Method

It is instructive to describe briefly the generation of $N + 1$ soliton solution from the N soliton solution using Darboux's procedure [27,28] which is closely related to Backlund transformation. As has been emphasized in the literature, the factorization method developed in connection with SUSYQM is a special case of Darboux construction. Darboux's method has been generalized by Crum [29] to the case of an arbitrary number of eigenfunctions.

The essence [30-34] of Darboux's method is to notice that if ϕ is a particular solution with an eigenvalue of the Schroedinger equation

$$H\psi = \left[-\frac{1}{2}\frac{d^2}{dx^2} + V(x) \right]\psi = E\psi \tag{7.38}$$

then the general solution of another Schroedinger equation

$$\mathcal{H}\Psi \equiv \left[-\frac{1}{2}\frac{d^2}{dx^2} + U(x) \right]\Psi = E\Psi \tag{7.39}$$

with $E(\neq e)$ is given by

$$\Psi = \frac{1}{\phi}(\psi\phi' - \psi'\phi) \tag{7.40}$$

and

$$U(x) = V(x) - \frac{d^2}{dx^2}ln\phi \tag{7.41}$$

Equation (7.44) closely resembles the expression (2.84) between the partner potentials for SUSYQM. The relevance of Darboux's method to the factorization scheme is brought out by the fact that $V_+(V_-)$ act like $U(V)$ with the correspondence $\psi_0^+ \leftrightarrow \frac{1}{\phi}$.

To have a nonsingular U, ϕ ought not to be vanishing. This restricts $e \leq E_0$, E_0 being the ground state energy of H. If we assume the existence of a solution of (7.38) that satisfies $u \to 0$ as $x \to -\infty$ and is nonvanishing for finite $x \in \mathbb{R}$, then the general solution of (7.38) for the case when $E = e$ can be obtained as

$$\phi(x) = u(x)\left(K + \int_x \frac{dx}{u^2} \right) \tag{7.42}$$

where $K \in \mathbb{R}^+$. We note that the energy spectrum of $\mathcal{H}$ defined by (7.39) is identical to that of H except for the presence of the ground state eigenvalue e. The corresponding normalized eigenfunction is given by $\sqrt{K}/\phi$.

To illustrate how Darboux's transformation works [19] in relating N and $N + 1$ soliton solutions we identify the potentials $Q^{(N)}$ and $Q^{(N+1)}$ with the N and $N + 1$ soliton solutions, respectively, and keep in mind that an M soliton solution has M bound states. Then we have

$$-\frac{1}{2}\frac{d^2\psi_i}{dx^2} + Q^{(N)}\psi_i = E_i\psi_i, \quad i = 1, 2, \ldots, N \tag{7.43}$$

$$-\frac{1}{2}\frac{d^2\Psi_i}{dx^2} + Q^{(N+1)}\Psi_i = E_i\Psi_i, \quad i = 1, 2, \ldots, N, N+1 \tag{7.44}$$

with the usual boundary conditions imposd namely ψ_i and $\Psi_i \to 0$ as $|x| \to \infty$. On the eigenvalues E_i we can, without losing generality, provide the ordering $E_{N+1} < E_N < \ldots < E_2 < E_1 < 0$.

We now supplement to (7.43) the Schroedinger equation for ψ_{N+1}. The latter, of course, cannot be a bounded solution

$$-\frac{1}{2}\frac{d^2\psi_{N+1}}{dx^2} + Q^{(N)}\psi_{N+1} = E_{N+1}\psi_{N+1} \tag{7.45}$$

To put (7.45) to use let us set $E_0 = E_{N+1}, V = Q^{(N)}, U = Q^{(N+1)}$ and identify the particular solution ϕ to be Ψ_{N+1}. Then $\Psi_{N+1} = \frac{1}{\psi_{N+1}}$ and we get from (7.41)

$$Q^{(N+1)} = Q^{(N)} + \frac{d^2}{dx^2}ln\psi_{N+1} \tag{7.46}$$

To get an $(N+1)$ solution the strategy is simple. We solve (7.45) for an unbounded solution corresponding to a given $Q^{(N)}$. This gives ψ_{N+1} which when substituted in (7.46) determines Q^{N+1}. In this way we avoid solving (7.44). The following example serves to make the point clear.

Set $E_i = -\lambda_i^2 < 0$ and start with the simplest vacuum case $V^{(0)} = 0$. Then from (7.45)

$$N = 0: -\frac{1}{2}\frac{d^2\psi_1}{dx^2} + \lambda_1^2\psi_1 = 0 \tag{7.47}$$

The unbounded solution is given by

$$\psi_1 = c_1 e^{\sqrt{2}\lambda_1 x} + c_2 e^{-\sqrt{2}\lambda_1 x} \tag{7.48}$$

where c_1, c_2 are arbitrary constants. Equation (7.46) will imply

$$Q^{(1)} = \frac{d^2}{dx^2}\log\left(c_1 e^{\sqrt{2}\lambda_1 x} + c_2 e^{-\sqrt{2}\lambda_2 x}\right) \tag{7.49}$$

Next consider the case $N = 1$ which reads from (7.45)

$$-\frac{1}{2}\frac{d^2\psi_2}{dx^2} + Q^{(1)}\psi_2 = -\lambda_2^2\psi_2 \qquad (7.50)$$

Equation (7.50) may be solved for an unbounded solution which turns out to be

$$\psi_2 = K\left(c_3 e^{\lambda_+ x} + c_4 e^{-\lambda_- x} + A\frac{\lambda_+}{\lambda_-}e^{-\lambda_- x} + B\frac{\lambda_-}{\lambda_+}e^{-\lambda_+ x}\right)/\psi_1 \quad (7.51)$$

where

$$\sqrt{2}(\lambda_1 \pm \lambda_2) = \lambda_\pm \qquad (7.52)$$
$$A = (c_2 c_3)/c_1 \qquad (7.53)$$
$$B = c_2 c_4/c_1 \qquad (7.54)$$

ψ_1 is given by (7.48) and K is a constant. In this way the hierarchy of potentials $Q^{(3)}, Q^{(4)}, \ldots$ can be determined.

7.6 SUSY and Conservation Laws in the KdV-SG Systems

Supersymmetric transformations can also be used [35] to connect the conserved quantities of the KdV to those of the SG equation. As stated earlier while the KdV equation arises in the context of water wave problems, a natural place for the SG equation to exist is in the motion of a closed string under an external field [36]. However, the SG equation can also be recognized as the evolution equation for a scalar field in $1 + 1$ dimensions in the presence of highly nonlinear self-interactions.

A remarkable property of the Lax form for the SG equation is that it is endowed with a supersymmetric structure corresponding to the eigenvalue problem of the L operator. At first sight it is not obvious why a link should exist between the KdV which is a nonrelativistic equation and SG which is a relativistic one. However what emerges is that supersymmetric transformations do not map the SG equation as a whole into the KdV. Only the eigenvalue equation of the L operator and as a consequence the conserved quantities for the two are transformed to each other.

Consider the spectral problem for the SG equation, which reads

$$L\chi = \xi\chi \tag{7.55}$$

where ξ is a constant, L is given by (7.33), and χ is the column matrix

$$\chi = \begin{pmatrix} \chi_1 \\ \chi_2 \end{pmatrix} \tag{7.56}$$

Then (7.55) translates to

$$2\chi_1' - i\psi'\chi_2 = \xi\chi_1 \tag{7.57a}$$

$$i\psi'\chi_1 - 2\chi_2' = \xi\chi_2 \tag{7.57b}$$

Defining the quantities $\chi_1 = \chi_1 \pm \chi_2$, (7.57a) and (7.57b) can be put in the form

$$-\frac{1}{2}\chi_+'' + \frac{1}{2}\left(W^2 - W'\right)\chi_+ = -\frac{\xi^2}{8}\chi_+ \tag{7.58a}$$

$$-\frac{1}{2}\chi_-'' + \frac{1}{2}\left(W^2 + W'\right)\chi_- = -\frac{\xi^2}{8}\chi_- \tag{7.58b}$$

with $W = -\frac{i\psi'}{2}$. The supersymmetric Hamiltonian H_s which acts on the two-component column $\begin{pmatrix} \chi_- \\ \chi_+ \end{pmatrix}$, may be expressed as $H_s =$ diag $H_-, H_+)$ similar to (2.28) where $H_\pm = -\frac{1}{2}\frac{d^2}{dx^2} \pm V_\pm$ and $V_\pm$ denoting the combinations $\frac{1}{2}\left(W^2 \mp W'\right)$. We thus see that a natural embedding of SUSY to be present in the eigenvalue problem of the L operator for the SG equation.

It is straightforward to relate the conservation laws in the KdV and SG systems. The eigenvalue problem of the L operator for the KdV system being given by (7.34), it follows that either of Equations (7.58) is identical to (7.34) if u transforms as

$$u_\pm \rightarrow \frac{1}{2}\left(W^2 \mp W'\right) \tag{7.59a}$$

$$\lambda \rightarrow -(\xi^2/4) \tag{7.59b}$$

Effectively, this implies that given the set of conserved quantities I_0, I_1, I_2, etc. of the KdV system, the corresponding ones for the SG system may be obtained through the mapping (7.59) and using the relation between W and ψ' given earlier. Indeed this turns out to be so (upto an overall constant in the coefficients of the integrals) as can be verified from (7.25) and comparing the transformed quantities with (7.27). It should be emphasized that the I's of SG equation do not depend on the arbitrariness of the sign in (7.59a), (7.21) being invariant under $\psi \to -\psi$.

Let us distinguish the two cases in (7.59) by u_+ and u_-. Our discussion in the preceding section now tells us that if suitable boundary conditions are prescribed to ψ then u_+ may be interpreted as an $(N+1)$ soliton solution if u_- corresponds to the N soliton solution. A relation between u_+ and u_- can be obtained on eliminating W

$$\left[(u_+ + u_-)^{1/2}\right]' = (u_- - u_+) \tag{7.60}$$

7.7 Supersymmetric KdV

In recent times a number of attempts have been made to seek supersymmetric extensions of the KdV equation [37-39]. This may be done in a superspace formalism (discussed in Chapter 2) replacing the coordinate x by the set (x, θ), θ being Gransmannian ($N = 1$ SUSY).

Let us consider a superfield Φ given by

$$\Phi = \xi(x) + \theta u(x) \tag{7.61}$$

where ξ is anticommuting in nature. Then the character of such a Φ is fermionic. Further, we define the covariant derivative to be

$$D = \theta \partial_x + \partial_\theta \tag{7.62}$$

The anticommuting nature of ξ and θ implies

$$
\begin{aligned}
D^2 &= \partial_x \\
\{D, \theta\} &= 0
\end{aligned}
\tag{7.63}
$$

where Q is the supersymmetric generator

$$Q = \partial_\theta - \theta \partial_x \tag{7.64}$$

Note that the supersymmetric transformations are realized according to

$$
\begin{aligned}
x &\rightarrow\ x - \eta\theta \\
\theta &\rightarrow\ \theta + \eta
\end{aligned}
\tag{7.65}
$$

where η is anticommuting. Indeed if we write

$$
\begin{aligned}
\delta\Phi &=\ \eta Q\Phi \\
&=\ \eta u + \theta\eta\xi'
\end{aligned}
\tag{7.66}
$$

we can deduce from (7.61)

$$
\begin{aligned}
\delta\xi &=\ \eta u(x) \\
\delta u &=\ \eta\xi'(x)
\end{aligned}
\tag{7.67}
$$

To have invariance under supersymmetric transformations (7.67) we need to work with quantities involving the covariant derivative and the superfield. So, in terms of these, the D and Φ, we can construct a supersymmetric version of the KdV equation, called the SKdV equation.

To derive SKdV, we observe that by multiplying both sides of (7.2) by the anticommuting variable θ we get the form

$$
\theta u_t = -\theta u''' + 6\theta u u'
\tag{7.68}
$$

Hence we can think of transforming $\theta u_t \rightarrow \Phi_t$ since Φ contains a θu part

$$
\Phi_t = \xi_t + \theta u_t
\tag{7.69}
$$

We also have the relations

$$
\begin{aligned}
D\Phi &=\ \theta\xi_x + u \\
D^2\Phi &=\ \theta u_x + \xi_x \\
D^4\Phi &=\ \theta u_{xx} + \xi_{xx} \\
D^6\Phi &=\ \theta u_{xxx} + \xi_{xxx}
\end{aligned}
\tag{7.70}
$$

So the dispersive term of (7.68) is seen to reside in $D^6\Phi$. However, the nonlinear term can be traced in the following two expressions

$$
\begin{aligned}
(D\Phi)(D^2\Phi) &=\ \theta\xi_x^2 + u\xi_x + \theta u u_x \\
D^2(\Phi D\Phi) &=\ -\theta\xi_x^2 - \theta\xi\xi_{xx} + u_x\xi + u\xi_x + 2\theta u u_x
\end{aligned}
\tag{7.71}
\tag{7.72}
$$

We therefore write the SKdV equation in a one-parameter representation

$$\Phi_t = -D^6\Phi + aD^2(\Phi D\Phi) + (6 - 2a)D\Phi D^2\Phi \qquad (7.73)$$

where a is arbitrary. We need to mention here that had we considered a bosonic extension of the KdV induced by a superfield $\chi = u(x) + \theta\alpha(x)$ in place of (7.61), the linearity of the fermionic field $\alpha(x)$ would have ensured that the resulting system yields the trivial case that KdV is self-generate with no influence from the quantity $\alpha(x)$.

However, (7.73) is nontrivial. We get, componentwise, for $u(x)$ and $\xi(x)$, the evolutions

$$
\begin{aligned}
u_t &= -u_{xxx} + 6uu_x - a\xi\xi_{xx} & (7.74)\\
\xi_t &= -\xi_{xxx} + au_x\xi + (6 - a)u\xi_x & (7.75)
\end{aligned}
$$

Thus the process of supersymmetrization affects the KdV equation in that it appears coupled with the ξ field.

Kupershmidt [38] also obtained a coupled set of equations involving the KdV

$$
\begin{aligned}
u_t &= -u_{xxx} + 6uu_x - 3\xi\xi_{xx} & (7.76)\\
\xi_t &= -4\xi_{xxx} + 6\xi_x u + 3\xi u_x & (7.77)
\end{aligned}
$$

and showed that superconformal algebra is related to it. But the above equations are not invariant under supersymmetric transformations.

We now turn to the Hamiltonian structure of the SKdV equation (7.73). In this regard, let us first of all demonstrate that the KdV is a bi-Hamiltonian system. We have already given the associated Hamiltonian structures for the conservation laws of the KdV equation in (7.25). Such a correspondence implies that the conserved quantities are in effect a sequence of Hamiltonians each generating its own evolution equation [40,41].

We now note that the KdV equation (7.2) can be expressed in two equivalent ways as follows $\left(D \equiv \frac{\partial}{\partial x}\right)$

$$\frac{\partial u}{\partial t} = Ou, \quad O = -D^3 + 2(Du + uD) \qquad (7.78)$$

and

$$\frac{\partial u}{\partial t} = D\left(3u^2 - D^2 u\right) \tag{7.79}$$

Introducing the functional forms of the Poisson bracket

$$\{A[u], B[u]\}_1 = \int d\tau \frac{\delta A[u]}{\delta u(\tau)} O \frac{\delta B[u]}{\delta u(\tau)} \tag{7.80}$$

$$\{A[u], B[u]\}_2 = \int d\tau \frac{\delta A[u]}{\delta u(\tau)} D \frac{\delta B[u]}{\delta u(\tau)} \tag{7.81}$$

and using the standard definitions of functional differentiation [42]

$$\frac{\delta F[X]}{\delta X(y)} = \lim_{\epsilon \to 0} \frac{1}{\epsilon} \left\{ F[\overline{X}] - F[X] \right\} \tag{7.82}$$

where

$$\overline{X}(x) = X(x) + \epsilon \delta(x - y) \tag{7.83}$$

it follows that

$$\{u(x), u(y)\}_1 = -\delta'''(x - y) + 4u\delta'(x - y)$$
$$+ 2u'\delta(x - y) \tag{7.84}$$
$$\{u(x), u(y)\}_2 = \delta'(x - y) \tag{7.85}$$

The definitions (7.80) and (7.81) then lead to

$$\frac{\partial u}{\partial t} = \{u(x), I_1\}_1$$
$$= \text{right hand side of (7.2)} \tag{7.86}$$

and

$$\frac{\partial u}{\partial t} = \{u(x), I_2\}_2$$
$$= \text{right hand side of (7.2)} \tag{7.87}$$

That the KdV equation is bi-Hamiltonian follows from the fact that it can be written in the form

$$u_t = \mathcal{D}^{(i)} \frac{\delta I_i}{\delta u} \tag{7.88}$$

in two different ways

$$\mathcal{D}^{(1)} = -D^3 + 2(Du + uD), \quad I_1 = \int dx u^2 \qquad (7.89)$$

$$\mathcal{D}^{(2)} = D, \quad I_2 = \int dx \left(u^3 + \frac{1}{2} u_x^2 \right) \qquad (7.90)$$

Given the constants of motion for the KdV summarized in (7.25), it is easy to arrive at their supersymmetric counterparts

$$I_1 \to J_1 = \frac{1}{2} \int dx d\theta \; \Phi D\Phi \qquad (7.91)$$

$$I_2 \to J_2 = \frac{1}{2} \int dx d\theta \; \left[2\Phi(D\Phi)^2 + D^2\Phi D^3\Phi \right] \qquad (7.92)$$

etc.

Consider J_2. In Poisson bracket formulation

$$\Phi_t = \{\Phi, J_2\} = -D^6\Phi + 4D\Phi D^2\Phi + 2\Phi D^3\Phi \qquad (7.93)$$

This picks out $a = 2$ from the SKdV equation (7.73). While deriving the above form we have made use of the following definition of the Poisson bracket

$$\{\Phi(x_1, \theta_1), \Phi(x_2, \theta_2)\} = \left(\theta_1 \frac{\partial}{\partial x_1} + \frac{\partial}{\partial \theta_1} \right) \Delta \qquad (7.94)$$

where

$$\Delta \equiv (\theta_2 - \theta_1)\delta(x_2 - x_1) \qquad (7.95)$$

Note that

$$\int dx_1 d\theta_1 F(x_1, \theta_1) \Delta = \int d\theta_1 (\theta_2 - \theta_1) \int dx_1 F(x_1, \theta_1)\delta(x_2 - x_1)$$

$$= \int d\theta_1 (\theta_2 - \theta_1) F(x_2, \theta_1)$$

$$= F(x_2, \theta_2) \qquad (7.96)$$

Inserting $\Phi(x_i, \theta_i) = \xi_i + \theta_i u_i (i = 1, 2)$ in (7.94), it follows that

$$\{\xi_1, \xi_2\} = -\delta(x_2 - x_1)$$

$$\{u_1, u_2\} = \frac{\partial}{\partial x_1} [\delta(x_2 - x_1)]$$

$$\{\xi_1, u_2\} = 0$$

$$\{u_1, \xi_2\} = 0 \qquad (7.97)$$

As such we can write

$$\partial_t \Phi = \mathcal{D} \frac{\delta H}{\delta \Phi} \tag{7.98}$$

where $\mathcal{D}$ is diagonal corresponding to the first two equations of (7.97). By expanding the right hand side of (7.92) in terms of quantities independent of θ and those depending on θ and calculating $\frac{\delta H}{\delta u}$ along with $\frac{\delta H}{\delta \xi}$ and using (7.82), the set of coupled equations (7.74) and (7.75) are seen to follow.

Corresponding to J_1, equivalence with (7.73) can be established for a different value of a namely $a = 3$ (see [39]). We therefore conclude that the supersymmetric KdV system does not have a bi-Hamiltonian structure. Actually, as pointed out by several authors, SKdV can be given a local meaning only [43-46].

The KdV equation has been extensively studied in relation to its integrability. Links with Virasoro algebra have been established through its second Hamiltonian structure [47]. Further the superconformal algebra was found to be related to supersymmetric extension of the KdV which is integrable [48]. A curious result has also been obtained concerning a pair of integrable fermionic extensions of the KdV equation : while one is bi-Hamiltonian but not supersymmetric, the other turns out to be supersymmetric (the one addressed to in this section) but not bi-Hamiltonian.

7.8   Conclusion

In this chapter we have discussed the role of SUSY in some nonlinear equations such as the KdV, MKdV, and SG. In the literature SUSY has also been used to study several aspects of nonlinear systems. More recently, hierarchy of lower KdV equations has been determined [50-53] which arise as a necessary part of supersymmetric constructions. It is now known that the supersymmetric structure of KdV and MKdV hierarchies leads to lower KdV equations and it becomes imperative to consider Miura's transformation in supersymmetric form. A supersymmetric structure has also been found to hold in Kadomtsev-Petviashvili (KP) hierarchies. In this connection it is relevant to mention that among various nonlinear evolution equations, the KdV-MKdV, KP and its modified partner are gauge equivalent to one another with the generating function coinciding with the equation for the corresponding gauge function [54]. Finally, we have discussed the SKdV equation and commented upon its Hamiltonian structures.

7.9 References

[1] D.J. Korteweg and G. de Vries, *Phil. Mag.*, **39**, 422, 1895.

[2] R. Rajaraman, *Solitons and Instantons*. North-Holland Publishing, Amsterdam, 1982.

[3] J. Scott-Russell, Rep. 14th Meeting British Assoc Adv. Sci., 311, John Murray, London, 1844.

[4] C.S. Gardner and G.K. Morikawa, *Courant Inst. Math. Sc. Res. Rep.*, NYO-9082, 1960.

[5] N.J. Zabusky, *Mathematical Models in Physical Sciences*, S. Drobot, Ed., Prentice-Hall, Englewood Cliffs, NJ, 1963.

[6] N.J. Zabusky, *J. Phys. Soc.*, **26**, 196, 1969.

[7] M.D. Kruskal, Proc IBM Scientific Computing Symposium on Large-scale Problems in Physics (IBM Data Processing Division, NY, 1965) p. 43.

[8] E. Fermi, J.R. Pasta, and S.M. Ulam, Los Alamos. Report Number LA-1940, Los Alamos, 1955.

[9] L. van Wijngaarden, *J. Fluid Mech.*, **33**, 465, 1968.

[10] M.C. Shen, *Siam J. Appl. Math.*, **17**, 260, 1969.

[11] C.S. Gardner, J.M. Greene, M.D. Kruskal, and R.M. Miura, *Phys. Rev. Lett.*, **19**, 1095, 1967.

[12] C.S. Gardner, J.M. Greene, M.D. Kruskal, and R.M. Miura, *Comm. Pure Appl. Math.*, **27**, 97, 1974.

[13] P.D. Lax, *Comm. Pure Appl. Math.*, **21**, 467, 1968.

[14] V.E. Zakharov and A.B. Shabat, *Func. Anal. Appl.*, **8**, 226, 1974.

[15] A.V. Bäcklund, *Math. Ann.*, **19**, 387, 1882.

[16] G.L. Lamb Jr., *J. Math. Phys.*, **15**, 2157, 1974.

[17] R.M. Miura, C.S. Gardner, and M.D. Kruskal, *J. Math. Phys.*, **9**, 1204, 1968.

[18] H.C. Morris, *J. Math. Phys.*, **18**, 530, 1977.

[19] M. Wadati, H. Sanuki, and K. Konno, *Prog. Theor. Phys.*, **53**, 419, 1975.

[20] A. Chodos, *Phys. Rev.*, **D21**, 2818, 1980.

[21] F. Gesztesy and H. Holden, *Rev. Math. Phys.*, **6**, 51, 1994.

[22] W. Kwong and J.L. Rosner, *Prog. Theor. Phys.*, Supplement, **86**, 366, 1986.

[23] Q. Wang, U.P. Sukhatme, W-Y Keung, and T.D. Imbo, *Mod. Phys. Lett.*, **A5**, 525, 1990.

[24] B. Bagchi, *Int. J. Mod. Phys.*, **A5**, 1763, 1990.

[25] A.A. Stahlhofen and A.J. Schramm, *Phys. Scripta.*, **43**, 553, 1991.

[26] J. Hruby, *J. Phys. A. Math. Gen.*, **22**, 1802, 1989.

[27] G. Darboux, *C.R. Acad Sci. Paris*, **92**, 1456, 1882.

[28] V.G. Bagrov and B.F. Samsonov, *Phys. Part Nucl.*, **28**, 374, 1997.

[29] M.M. Crum, *Quart J. Math.*, **6**, 121, 1955.

[30] D.L. Pursey, *Phys. Rev.*, **D33**, 1048, 1986.

[31] D.L. Pursey, *Phys. Rev.*, **D33**, 2267, 1986.

[32] M.M. Nieto, *Phys. Lett.*, **B145**, 208, 1984.

[33] I.M. Gel'fand and B.M. Levitan, *Am. Math. Soc. Trans.*, **1**, 253, 1955.

[34] P.B. Abraham and H.E. Moses, *Phys. Rev.*, **A22**, 1333, 1980.

[35] B. Bagchi, A. Lahiri, and P.K. Roy, *Phys. Rev.*, **D39**, 1186, 1989.

[36] F. Lund and T. Regge, *Phys. Rev.*, **D14**, 1524, 1976.

[37] Yu I. Manin and A.O. Radul, *Comm. Math. Phys.*, **98**, 65, 1985.

[38] B.A. Kupershmidt, *Phys. Lett.*, **A102**, 213, 1984.

[39] P. Mathieu, *J. Math. Phys.*, **29**, 2499, 1988.

[40] A. Das, *Phys. Lett.*, **B207**, 429, 1988.

[41] Q. Liu, *Lett. Math. Phys.*, **35**, 115, 1995.

[42] R.J. Rivers, *Path. Integral Methods in Quantum Field Theory*, Cambridge University Press, Cambridge, 1987.

[43] T. Inami and H. Kanno, *Comm. Math. Phys.*, **136**, 519, 1991.

[44] W. Oevel and Z. Poponicz, *Comm. Math. Phys.*, **139**, 441, 1991.

[45] J. Figueroa-O' Farill, J. Mas, and E. Rames, *Rev. Math. Phys.*, **3**, 479, 1991.

[46] J.C. Brunelli and A. Das, *Phys. Lett.*, **B337**, 303, 1994.

[47] J.L. Gervais and A. Neveu, *Nucl. Phys.*, **B209**, 125, 1982.

[48] P. Mathieu, *Phys. Lett.*, **B203**, 287, 1988.

[49] P. Mathieu, *Phys. Lett.*, **B208**, 101, 1988.

[50] V.A. Andreev and M.V. Burova, *Theor. Math. Phys.*, **85**, 376, 1990.

[51] V.A. Andreev and M.V. Shmakova, *J. Math. Phys.*, **34**, 3491, 1993.

[52] A.V. Samohin, *Sov. Math. Dokl.*, **21**, 93, 1980.

[53] A.V. Samohin, *Sov. Math. Dokl.*, **25**, 56, 1982.

[54] B.G. Konopelchenko, *Rev. Math. Phys.*, **2**, 339, 1990.

CHAPTER 8

Parasupersymmetry

8.1 Introduction

In the literature SUSYQM has been extended [1-16] to what constitutes parasupersymmetric quantum mechanics (PSUSYQM). To understand its underpinnings we first must note that the fermionic operators a and a^+ obeying (2.12) and (2.13) also satisfy the commutation condition

$$[a^+, a] = \text{diag}\left[\frac{1}{2}, -\frac{1}{2}\right] \tag{8.1}$$

The entries in the parenthesis of the right-hand-side can be looked upon as the eigenvalues of the 3rd component of the spin $\frac{1}{2}$ operator.

The essence of a minimal (that is of order $p = 2$) PSUSYQM scheme is to replace the right-hand-side of (8.1) by the eigenvalues of the 3rd component of the spin 1 operator. Thus in $p = 2$ PSUSYQM a new set of operators c and c^+ is introduced with the requirement

$$[c^+, c] = 2 \, \text{diag}(1, 0, -1) \tag{8.2}$$

A plausible set of representations for c and c^+ satisfying (8.2) is given by the matrices

$$c = \sqrt{2}\begin{pmatrix} 0 & 0 & 0 \\ 1 & 0 & 0 \\ 0 & 1 & 0 \end{pmatrix}$$

$$c^+ = \sqrt{2} \begin{pmatrix} 0 & 1 & 0 \\ 0 & 0 & 1 \\ 0 & 0 & 0 \end{pmatrix} \tag{8.3}$$

From the above it is clear that the nature of the operators c and c^+ is parafermionic [17-21] of order 2

$$c^3 = c^{+^3} = 0$$
$$cc^+c = 2c$$

$$c^2c^+ + c^+c^2 = 2c \tag{8.4a, b, c}$$

Note that (8.4b) and (8.4c) are also consistent with the alternative algebra in terms of double commutators

$$[c, [c, c^+]] = -2c,$$
$$[c^+, [c, c^+]] = 2c^+ \tag{8.5}$$

One is thus motivated into defining a set of parasupersymmetric charges by combining the usual bosonic ones with parafermionic operators. At the level of order 2 these are given by

$$Q = b \otimes c^+$$
$$Q^+ = b^+ \otimes c \tag{8.6}$$

which generalize the supersymmetric forms (2.21). In the following, we shall assume the same notations for the parasupercharges as for the supercharges.

With (8.4a) holding, it is obvious that

$$Q^3 = 0 = Q^{+^3} \tag{8.7}$$

More generally one can construct parasupercharges of order p with the properties

$$(Q)^{p+1} = (Q^+)^{p+1} = 0, \ p = 1, 2, \ldots \tag{8.8}$$

When $p = 1$, (8.8) implies the usual nilpotency conditions of the supercharges in SUSYQM. Note that for higher order ($p > 2$) PSUSY, the diagonal term in the right-hand-side of (8.2) needs to be replaced by the generalized matrix of the type diag $\left(\frac{p}{2}, \frac{p}{2} - 1, \ldots, -\frac{p}{2} + 1, -\frac{p}{2}\right)$.

8.2   Models of PSUSYQM

(a) The scheme of Rubakov and Spiridonov

Rubakov and Spiridonov [1] were the first to propose a generalization of the Witten supersymmetric Hamiltonian (2.22) to a PSUSY form. They defined the PSUSY Hamiltonian H_p as arising from the relations

$$Q^2Q^+ + QQ^+Q + Q^+Q^2 = 2QH_p \tag{8.9}$$
$$Q^{+2}Q + Q^+QQ^+ + QQ^{+2} = 2Q^+H_p \tag{8.10}$$

where Q and Q^+ apart from obeying (8.7) also commute with H_p

$$[H_p, Q] = 0 = [H_p, Q^+] \tag{8.11}$$

Keeping in mind the transitions (2.34), the parasupercharges Q and Q^+ can be given a matrix representation as follows

$$(Q)_{ij} = \frac{1}{\sqrt{2}} \left[\frac{d}{dx} + W_i(x) \right] \delta_{i+1,j}, \ i,j = 1,2,3 \tag{8.12a}$$

That is

$$Q = \frac{1}{\sqrt{2}} \begin{pmatrix} 0 & A_1^+ & 0 \\ 0 & 0 & A_2^+ \\ 0 & 0 & 0 \end{pmatrix} \tag{8.12b}$$

Also

$$(Q^+)_{ij} = \frac{1}{\sqrt{2}} \left[-\frac{d}{dx} + W_j(x) \right] \delta_{i,j+1}, \ i,j = 1,2,3 \tag{8.12c}$$

That is

$$Q^+ = \frac{1}{\sqrt{2}} \begin{pmatrix} 0 & 0 & 0 \\ A_1^- & 0 & 0 \\ 0 & A_2^- & 0 \end{pmatrix} \tag{8.12d}$$

In (8.12b) and (8.12d) the notations $A_i^{\pm}$ ($i = 1,2$) stand for

$$A_i^{\pm} = W_i(x) \pm \frac{d}{dx} \tag{8.13}$$

Here the parasupercharges Q and Q^+ are defined in terms of a pair of superpotentials $W_1(x)$ and $W_2(x)$ indicating a switchover from the order $p = 1$ (which is SUSYQM) to $p = 2$ (which is PSUSYQM of order 2).

Given (8.12), the PSUSY algebra (8.9) and (8.10) lead to the following diagonal form for H_p

$$H_p = \text{diag}\,(H_1, H_2, H_3) \tag{8.14}$$

with $H_1, H_2,$ and H_3 in terms of $A_i^{\pm}(i = 1, 2)$ being

$$A_1^- H_1 = \frac{1}{4}\left(A_1^- A_1^+ + A_2^+ A_2^-\right) A_1^-$$

$$H_2 = \frac{1}{4}\left(A_1^- A_1^+ + A_2^+ A_2^-\right)$$

$$A_2^+ H_3 = \frac{1}{4}\left(A_1^- A_1^+ + A_2^+ A_3^-\right) A_2^+ \tag{8.15}$$

Now, the above representations are of little use unless H_1 and H_3, like H_2, are reducible to tractable forms. One way to achieve this is to impose upon (8.15) the constraint

$$A_1^- A_1^+ = A_2^+ A_2^- + c \tag{8.16}$$

where c is a constant. However, the components of H_p can be given expressions which are independent of c, namely,

$$H_1 = \frac{1}{4}\left(-2\frac{d^2}{dx^2} + W_1^2 + W_2^2 + 3W_1' + W_2'\right)$$

$$H_2 = \frac{1}{4}\left(-2\frac{d^2}{dx^2} + W_1^2 + W_2^2 - W_1' + W_2'\right)$$

$$H_3 = \frac{1}{4}\left(-2\frac{d^2}{dx^2} + W_1^2 + W_2^2 - W_1' - 3W_2'\right) \tag{8.17}$$

where the functions W_1 and W_2, on account of (8.16), are restricted by

$$W_2^2 - W_1^2 + W_1' + W_2' + c = 0 \tag{8.18}$$

Let us examine the particular case when the derivatives of the superpotentials W_1 and W_2 are equal

$$W_1' = W_2' \tag{8.19}$$

The above proposition leads to a simple form of the PSUSY Hamiltonian H_p which affords a straightforward physical meaning

$$H_p = \left[-\frac{1}{2}\frac{d^2}{dx^2} + V(x) \right] \mathbb{1}_3 + B(x) J_3 \qquad (8.20)$$

where

$$
\begin{aligned}
V(x) &= \frac{1}{4}\left(W_1^2 + W_2^2\right)\\[4pt]
\mathbb{1}_3 &= \begin{pmatrix} 1 & 0 & 0 \\ 0 & 1 & 0 \\ 0 & 0 & 1 \end{pmatrix}\\[4pt]
B(x) &= \frac{dW_1}{dx}\\[4pt]
J_3 &= \begin{pmatrix} 1 & 0 & 0 \\ 0 & 0 & 0 \\ 0 & 0 & -1 \end{pmatrix}
\end{aligned}
\qquad (8.21)
$$

The interpretation of H_p is self-evident, it represents the motion of a spin 1 particle placed in a magnetic field $\vec{B}$ directed along the 3rd axis.

Two solutions of (8.18) corresponding to (8.19) are of the types either

$$W_1 = W_2 = \omega_1 x + \omega_2 \qquad (8.22)$$

or

$$W_1 = \omega_1 e^{-kx} + \omega_2, \; W_2 = \omega_1 + k \qquad (8.23)$$

while the first one is for the homogeneous magnetic field, the second one is for the inhomogeneous type.

The above PSUSY scheme corresponding to the supercharges given by (8.12) can also be put in an alternative form by making use of the fact that at the level of order 2 there are 2 independent para-supercharges. Actually we can write down Q as a linear combination of 2 supercharges Q_1 and Q_2, namely,

$$
\begin{aligned}
Q &= Q_1 + Q_2\\
Q^+ &= Q_1^+ + Q_2^+
\end{aligned}
\qquad (8.24)
$$

where

$$Q_1 = \frac{1}{\sqrt{2}} \begin{pmatrix} 0 & A_1^+ & 0 \\ 0 & 0 & 0 \\ 0 & 0 & 0 \end{pmatrix}$$

$$Q_2 = \frac{1}{\sqrt{2}} \begin{pmatrix} 0 & 0 & 0 \\ 0 & 0 & A_2^+ \\ 0 & 0 & 0 \end{pmatrix}$$

$$Q_1^+ = \frac{1}{\sqrt{2}} \begin{pmatrix} 0 & 0 & 0 \\ A_1^- & 0 & 0 \\ 0 & 0 & 0 \end{pmatrix}$$

$$Q_2^+ = \frac{1}{\sqrt{2}} \begin{pmatrix} 0 & 0 & 0 \\ 0 & 0 & 0 \\ 0 & A_2^- & 0 \end{pmatrix} \tag{8.25}$$

It is easily verified that Q_i and Q_i^+ $(i = 1, 2)$ are in fact super-charges in the sense that they are endowed with the properties

$$Q_i^2 = 0 = Q_i^{+2} \quad i = 1, 2 \tag{8.26}$$

for $c = 0$. Besides, they satisfy

$$Q_i Q_j^+ = Q_i^+ Q_j = 0 \quad (i \neq j), \ i = 1, 2 \tag{8.27}$$

A more transparent account of the natural embedding of the supersymmetric algebra in (8.9) and (8.10) can be brought about by invoking the hermitean charges

$$Q_1 = \frac{1}{\sqrt{2}} (Q^+ + Q) \tag{8.28}$$

$$Q_2 = \frac{1}{\sqrt{2i}} (Q^+ - Q) \tag{8.29}$$

We may work out

$$\begin{aligned} Q_1^3 &= \frac{1}{2\sqrt{2}} (Q^+ + Q)(Q^+ + Q)(Q^+ + Q) \\ &= \frac{1}{2\sqrt{2}} \left(Q^{+2}Q + Q^+QQ^+ + Q^+Q^2 + QQ^{+2} + QQ^+Q + Q^2Q^+ \right) \\ &= \frac{1}{2\sqrt{2}} (2Q^+H_p + 2QH_p) \\ &= \frac{1}{\sqrt{2}} (Q + Q^+)H_p \\ &= Q_1 H_p \tag{8.30} \end{aligned}$$

and similarly

$$Q_2^3 = Q_2 H_p \qquad (8.31)$$

Further relations, (8.9) and (8.10), can be exploited to yield for the real and imaginary parts, the conditions

$$Q_1 Q_2^2 + Q_2 Q_1 Q_2 + Q_2^2 Q_1 = Q_1 H_p \qquad (8.32)$$
$$Q_2 Q_1^2 + Q_1 Q_2 Q_1 + Q_1^2 Q_2 = Q_2 H_p \qquad (8.33)$$

To see a connection to SUSY we write, say (8.33), in the following manner

$$[\{Q_1, Q_1\} - 2H_p] Q_2 + [\{Q_1, Q_2\} + \{Q_2, Q_1\}] Q_1 = 0 \qquad (8.34)$$

This suggests a combined relation $(i, j, k = 1, 2)$

$$[\{Q_i, Q_j\} - 2H_p\delta_{ij}] Q_k + [\{Q_j, Q_k\} - 2H_p\delta_{jk}] Q_i$$
$$+ [\{Q_k, Q_i\} - 2H_p\delta_{ki}] Q_j = 0 \ (8.35)$$

to be compared with the SUSY formula (2.44).

(b) The scheme of Beckers and Debergh

Beckers and Debergh [2] made an interesting observation that the choice of the Hamiltonian in defining a PSUSY system is not unique. They constructed a new Hamiltonian for PSUSY by requiring H_p to obey the following double commutator

$$[Q, [Q^+, Q]] = Q H_p \qquad (8.36)$$
$$[Q^+, [Q, Q^+]] = Q^+ H_p \qquad (8.37)$$

in addition to the obvious properties (8.7) and (8.11).

One is easily convinced that (8.36) and (8.37) are inequivalent to the corresponding ones (8.9) and (8.10) of Rubakov and Spiridonov. This follows from the fact that an equivalence results in the conditions

$$QQ^+Q = QH_p \qquad (8.38)$$
$$Q^2Q^+ + Q^+Q^2 = QH_p \qquad (8.39)$$

a feature which is not present in the model of Rubakov and Spiridonov. As such the parasuper Hamiltonian dictated by (8.36) and (8.37) is nontrivial and offers a new scheme of PSUSYQM.

To obtain a plausible representation of H_p such as in (8.14) we assume the parasupercharges Q and Q^+ to be given by the matrices (8.12b) and (8.12d) but controlled by the constraint.

$$W_2^2 - W_1^2 + (W_1' + W_2') = 0 \qquad (8.40)$$

in place of (8.18). Note that the latter differs from (8.40) in having just $c = 0$. However, the components of H_p are significantly different from those in (8.17). Here H_1, H_2, and H_3 are

$$H_1 = \frac{1}{2}\left(-\frac{d^2}{dx^2} + 2W_1^2 - W_2^2 - W_2'\right)$$

$$H_2 = \frac{1}{2}\left(-\frac{d^2}{dx^2} + 2W_2^2 - W_1^2 + 2W_2' + W_1'\right)$$

$$H_3 = \frac{1}{2}\left(-\frac{d^2}{dx^2} + 2W_2^2 - W_1^2 + W_1'\right) \qquad (8.41)$$

An interesting particular case stands for

$$W_1 = -W_2 = \omega x \qquad (8.42)$$

(ω is a constant) which is consistent with (8.40). It renders H_p to the form

$$H_p = \frac{1}{2}\left(-\frac{d^2}{dx^2} + \omega^2 x^2\right)\mathbb{1}_3 + \frac{\omega}{2}\ \mathrm{diag}(1, -1, 1) \qquad (8.43)$$

which can be looked upon as a natural generalization of the supersymmetric oscillator Hamiltonian.

In terms of the hermitean quantities Q_1 and Q_2, here one can derive the relations

$$\begin{aligned}
Q_1^3 &= Q_1 H_p \\
Q_2^3 &= Q_2 H_p \\
Q_1 Q_2 Q_1 &= Q_2 Q_1 Q_2 = 0 \\
Q_1^2 Q_2 + Q_2 Q_1^2 &= Q_2 H_p \\
Q_2^2 Q_1 + Q_1 Q_2^2 &= Q_1 H_p
\end{aligned} \qquad (8.44)$$

Although (8.44) are consistent with (8.30), (8.31), (8.32), and (8.33) of the Rubakov-Spiridonov scheme, the converse is not true.

Since $c = 0$ in the Beckers-Debergh model, the constraint equation (8.40) is an outcome of the operator relation

$$A_1^- A_1^+ = A_2^+ A_2^-$$ (8.45)

Because of (8.45), the component Hamiltonians in (8.41) are essentially the result of the following factorizations

$$H_1 = \frac{1}{2} A_1^+ A_1^-$$

$$H_2 = \frac{1}{2} A_1^- A_1^+$$

$$H_3 = \frac{1}{2} A_2^- A_2^+$$ (8.46)

We can thus associate with

$$H_p = \frac{1}{2} \begin{pmatrix} A_1^+ A_1^- & 0 & 0 \\ 0 & A_1^- A_1^+ & 0 \\ 0 & 0 & A_2^- A_2^+ \end{pmatrix}$$ (8.47)

two distinct supersymmetric Hamiltonians given by

$$H_s^{(1)} = \frac{1}{2} \begin{pmatrix} A_1^+ A_1^- & 0 \\ 0 & A_1^- A_1^+ \end{pmatrix}$$ (8.48a)

$$H_s^{(2)} = \frac{1}{2} \begin{pmatrix} A_2^+ A_2^- & 0 \\ 0 & A_2^- A_2^+ \end{pmatrix}$$ (8.48b)

This exposes the supersymmetric connection of the Beckers-Debergh Hamiltonian.

Let us now comment on the relevance of the parasupersymmetric matrix Hamiltonian in higher derivative supersymmetric schemes [22-24]. Indeed the representations (8.46) are strongly reminiscent of the components h_-, h_0, and h_+ given in Section 4.9 of Chapter 4 [see the remarks following (4.109b)]. Recall that we had expressed the quasi-Hamiltonian K as the square of the Schroedinger type operator h by setting the parameters $\mu = \lambda = 0$. The resulting components of h, namely, h_- and h_+, can be viewed as being derivable from the second-order PSUSY Hamiltonian H_p given by (8.47) by deleting its intermediate piece. Conversely we could get the $p = 2$ PSUSY form of the Hamiltonian from the components h_- and h_+ by glueing these together to form the (3×3) system (8.47). Thus

SSUSY Hamiltonian can be interpreted as being built up from two ordinary supersymmetric Hamiltonians such as of the types (8.48a) and (8.48b).

Consider the case when $c \neq 0$ (which is consistent with the Rubakov-Spiridonov scheme) corresponds to the situation (4.112) where $\nu \neq 0$. Here also glueing of supersymmetric Hamiltonians can be done to arrive at a 3×3 matrix structure. Truncation of the intermediate component then yields the SSUSY Hamiltonian in its usual 2×2 form.

8.3 PSUSY of Arbitrary Order p

Khare [25,26] has shown that a PSUSY model of arbitrary order p can be developed by generalizing the fundamental equations (8.7) and (8.9)-(8.10) to the forms ($p \geq 2$)

$$Q^{p+1} = 0 \qquad (8.49)$$
$$[H_p, Q] = 0 \qquad (8.50)$$

$$Q^p Q^+ + Q^{p-1} Q^+ Q + \ldots + Q^+ Q^p = p Q^{p-1} H_p \qquad (8.51)$$

along with their hermitean conjugated relations.

The parasupercharges Q and Q^+ can be chosen to be $(p+1) \times (p+1)$ matrices as natural extensions of the $p = 2$ scheme

$$(Q)_{\alpha\beta} = A_\alpha^+ \, \delta_{\alpha+1,\beta} \qquad (8.52)$$
$$(Q^+)_{\alpha\beta} = A_\beta^- \, \delta_{\alpha,\beta+1} \qquad (8.53)$$

where $\alpha, \beta = 1, 2, \ldots p + 1$. The generalized matrices for c and c^+ read

$$c = \begin{pmatrix} 0 & 0 & \cdots & 0 & 0 \\ \sqrt{p} & 0 & \cdots & 0 & 0 \\ 0 & \sqrt{2(p-1)} & \cdots & 0 & 0 \\ \vdots & \vdots & \vdots & \vdots & \vdots \\ 0 & 0 & \cdots & \sqrt{p} & 0 \end{pmatrix}$$

$$c^+ = \begin{pmatrix} 0 & \sqrt{p} & 0 & \cdots & 0 \\ 0 & 0 & \sqrt{2(p-1)} & \cdots & 0 \\ \cdots & \cdots & \cdots & \cdots & \cdots \\ \cdots & \cdots & \cdots & \cdots & \cdots \\ 0 & 0 & \cdots & \cdots & \sqrt{p} \\ 0 & 0 & \cdots & \cdots & 0 \end{pmatrix} \qquad (8.54)$$

In (8.52) and (8.53), A_α^+ and A_α^- can be written explicitly as

$$A_\alpha^+ = \frac{d}{dx} + W_\alpha(x) \qquad (8.55)$$

$$A_\alpha^- = -\frac{d}{dx} + W_\alpha(x), \quad \alpha = 1, 2, \ldots, p \qquad (8.56)$$

The Hamiltonian H_p being diagonal is given by

$$H_p = \text{diag}(H_1, H_2, \ldots, H_{p+1}) \qquad (8.57)$$

where

$$H_r = -\frac{1}{2}\frac{d^2}{dx^2} + \frac{1}{2}\left(W_r^2 + W_r'\right) + \frac{1}{2}c_r, \quad r = 1, 2, \ldots p \qquad (8.58)$$

$$H_{p+1} = -\frac{1}{2}\frac{d^2}{dx^2} + \frac{1}{2}\left(W_p^2 - W_p'\right) + \frac{1}{2}c_p \qquad (8.59)$$

subject to the constraint

$$W_{r-1}^2 - W_{r-1}' + c_{r-1} = W_r^2 + W_r' + c_r, \quad r = 2, 3, \ldots p \qquad (8.60)$$

The constants $c_1, c_2, \ldots, c_p$ are arbitrary and have the dimension of energy. However, if one explicitly works out the quantities $Q^p Q^+, Q^{p-1} Q^+ Q$, etc. appearing in (8.51), it turns out that the constants $c_1, c_2, \ldots c_p$ are not independent

$$c_1 + c_2 + \ldots + c_{p-1} + c_p = 0 \qquad (8.61)$$

Let us verify (8.58)-(8.60) for the case $p = 2$ which we have already addressed to in the Rubakov-Spiridonov scheme. First of all we notice that on taking derivatives of both sides, (8.60) matches with the corresponding derivative version of (8.18) for $p = 2$. Secondly, on using (8.60) and (8.61) we can express again (8.58) and (8.59) as

$$
\begin{aligned}
H_1 &= -\frac{1}{2}\frac{d^2}{dx^2} + \frac{1}{2p}\left(W_1^2 + W_2^2 + \ldots + W_p^2\right) + \left(1 - \frac{1}{2p}\right)W_1' \\
&\quad + \left(1 - \frac{3}{2p}\right)W_2' + \ldots + \frac{3}{2p}W_{p-1}' + \frac{1}{2p}W_p', \\
H_2 &= H_1 - W_1' \\
&\cdots \qquad \cdots\cdots \\
H_{r+1} &= H_r - W_r' \\
&\cdots \qquad \cdots\cdots \\
H_{p+1} &= H_p - W_p' \qquad\qquad\qquad\qquad\qquad\qquad (8.62)
\end{aligned}
$$

Now it is evident that (8.62) yields (8.17) on putting $p = 2$. Likewise, the Hamiltonians for higher order cases can be obtained on putting $p = 3, 4, \ldots$ etc.

The particular case when

$$
W_1 = W_2 = \ldots = W_p = \omega x \qquad\qquad (8.63)
$$

yields the PSUSY oscillator Hamiltonian which, of course, is realized in terms of bosons and parafermions of order p. This Hamiltonian can be written as

$$
\begin{aligned}
H &= \left[-\frac{1}{2}\frac{d^2}{dx^2} + \frac{1}{2}\omega^2 x^2\right]\|\! \\
&\quad - \omega\,\mathrm{diag}\left(\frac{p}{2}, \frac{p}{2} - 1, \ldots, -\frac{p}{2} + 1, -\frac{p}{2}\right), p \geq 2 \quad (8.64)
\end{aligned}
$$

whose spectrum is

$$
E_{n,\nu} = \left(n + \frac{1}{2} - \nu\right)\omega \qquad\qquad (8.65)
$$

where

$$
\begin{aligned}
n &= 0, 1, 2, \ldots \\
\nu &= \frac{p}{2}, \frac{p}{2} - 1, \ldots, -\frac{p}{2} \qquad\qquad (8.66)
\end{aligned}
$$

From (8.65) one can see that the ground state is nondegenerate with its energy given by

$$
E_{0,\frac{p}{2}} = \left(\frac{1}{2} - \frac{p}{2}\right)\omega \qquad\qquad (8.67)
$$

It is clearly negative. Further, the pth excited state is $(p+1)$-fold degenerate.

To conclude, there have been several variants of PSUSYQM suggested by different authors. In [27] a generalization of SUSYQM was considered leading to fractional SUSYQM having a structure similar to PSUSYQM. Durand et al. [16] discussed a conformally invariant PSUSY whose algebra involves the dilatation operator, the conformal operator, the hypercharge, and the superconformal charge. A bosonization of PSUSY of order two has also been explored [28]. In this regard, a realization of C_λ extended oscillator algebra was shown [29] to provide a bosonization of PSUSYQM of order $p = \lambda - 1$ for any λ.

8.4 Truncated Oscillator and PSUSYQM

The annihilation and creation operators of a normal bosonic oscillator subjected to the quantum condition $[b, b^+] = 1$ are well known to possess infinite dimensional matrix representations

$$
b = \begin{pmatrix}
0 & \sqrt{1} & 0 & 0 & \cdots \\
0 & 0 & \sqrt{2} & 0 & \cdots \\
0 & 0 & 0 & \sqrt{3} & \cdots \\
\vdots & \vdots & \vdots & \vdots & \ddots
\end{pmatrix}
$$

$$
b^+ = \begin{pmatrix}
0 & 0 & 0 & 0 & \cdots \\
\sqrt{1} & 0 & 0 & 0 & \cdots \\
0 & \sqrt{2} & 0 & 0 & \cdots \\
0 & 0 & \sqrt{3} & 0 & \cdots \\
\vdots & \vdots & \vdots & \vdots & \ddots
\end{pmatrix} \tag{8.68}
$$

A truncated oscillator is the one characterized [30,31] by some finite dimensional representations of (8.68). The interest in finite dimensional Hilbert space (FHS) comes from the recent developments [32,33] in quantum phase theory which deals with a quantized harmonic oscillator in a FHS and which finds important applications [34-36] in problems of quantum optics. In this section we discuss PSUSYQM when the PSUSY is between the normal bosons [described by the annihilation and creation operators b and b^+ obeying (8.68)] and those corresponding to a truncated harmonic oscillator which behave, as we shall presently see, like an exotic para Fermi oscillator.

The Hamiltonian of a truncated oscillator is given by

$$H = \frac{1}{2}\left(P^2 + Q^2\right) \tag{8.69}$$

where P and Q are the corresponding truncated versions of the canonical observables p and q of the standard harmonic oscillator.

The variables P and Q, however, are not canonical in that their commutation has the form

$$QP - PQ = i(I - NK) \tag{8.70}$$

where I and K are N dimensional matrices, I being the unit matrix and K having the form [37]

$$K = \begin{pmatrix} 0 & 0 & \ldots & 0 \\ 0 & 0 & \ldots & 0 \\ \vdots & \vdots & \vdots & \vdots \\ 0 & 0 & \ldots & 0 \\ 0 & 0 & \ldots & 1 \end{pmatrix} \tag{8.71}$$

In other words, K is a diagonal matrix with the last element unity as the only nonzero element. In (8.70), $N(> 1)$ is a parameter and signifies that at the N-th now and column the matrices p and q have been truncated.

We first show that K plays the role of a projection operator $|t><t|$ in the FHS which for concreteness is taken to be a $(t+1)$-dimensional Fock space

$$\mathcal{T} = \{|0>, |1>, \ldots |t>\} \tag{8.72}$$

t being a positive interger. This enables us to connect Buchdahl's work [30] on the truncated oscillator and some recent works which have tried to explain [32,33] the existence of a hermitean phase operator. We also bring to surface the remarkable similarities between the rules obeyed by the representative matrices of the truncated oscillator and those of the parafermionic operators.

(a) Truncated oscillator algebra

Let us begin by writing down the Hamiltonian for the linear harmonic oscillator

$$H = \frac{1}{2}\left(p^2 + q^2\right) \tag{8.73}$$

where the observables q and p satisfy $[q, p] = i\hbar$.

The associated lowering and raising operators b and b^+ are defined by (2.2). These obey, along with the bosonic number operator N_B, the relations

$$[N_B, b] = -b$$
$$[N_B, b^+] = b^+ \tag{8.74}$$

where $N_B = b^+ b$. The essential properties of b and b^+ may be summarized in terms of the following nonvanishing matrix elements

$$< n'|b|n > = \sqrt{n}\, \delta_{n',n-1}$$
$$< n'|b^+|n > = \sqrt{n+1}\, \delta_{n',n+1} \tag{8.75}$$

where $n, n' = 0, 1, 2, \ldots$. The above relations reflect the consequences of seeking matrix representations of b and b^+ or equivalently p and q.

Let us consider the truncation of the matrices (8.68) for b and b^+ at the Nth row and column and call the corresponding new operators to be B and B^+ where

$$B = \frac{1}{\sqrt{2}}(Q + iP)$$
$$B^+ = \frac{1}{\sqrt{2}}(Q - iP) \tag{8.76}$$

Using (8.70) the modified commutation relation for B and B^+ reads

$$[B, B^+] = 1 - NK \tag{8.77}$$

where

$$KB = 0$$
$$K^2 = K \neq 0 \tag{8.78}$$

When $N = 2$, a convenient set of representations for (8.77) is

$$
\begin{aligned}
B &= \frac{1}{2}\sigma_+ \\
B^+ &= \frac{1}{2}\sigma_- \\
K &= \frac{1}{2}(1 - \sigma_3) \quad\quad (8.79)
\end{aligned}
$$

where $\sigma_\pm$ have been already defined in Chapter 2.

Let us now suppose that a FHS is generated by an orthonormal set of kets $|i>, i = 0, 1, \ldots t$ along with a completeness condition, namely [38]

$$
\begin{aligned}
<j|k> &= \delta_{j,k} \\
\sum_{j=0}^{t} <j|j> &= I \quad\quad (8.80)
\end{aligned}
$$

The operators B and B^+ may be introduced through

$$
\begin{aligned}
B &= \sum_{m=0}^{t} \sqrt{m}|m-1><m| \\
B^+ &= \sum_{m=0}^{t} \sqrt{m}|m><m-1| \quad\quad (8.81)
\end{aligned}
$$

These ensure that

$$
\begin{aligned}
B|m> &= \sqrt{m}|m-1> \\
B|0> &= 0 \\
B^+|m> &= \sqrt{m+1}|m+1> \\
B^+|t> &= 0 \quad\quad (8.82a, b, c, d)
\end{aligned}
$$

where $m = 0, 1, 2, \ldots t$.

Any ket $|j>$ can be derived from the vacuum by applying B^+ on $|0>$ j times

$$
|j> = \frac{1}{\sqrt{j!}}(B^+)^j|0> \quad\quad (8.83)
$$

where $j = 0, 1, 2, \ldots t$. However since the dimension of the Hilbert space is finite, the ket $|t>$ cannot be pushed up to a higher position.

This is expressed by (8.82d). Note that the Fredholm index δ vanishes in this case [see (4.41b)].

In view of the conditions (8.80), the expansions (8.81) imply that

$$[B, B^+] = 1 - (t + 1)|t ><t| \tag{8.84}$$

By comparing with (8.77) it is clear that K plays the role of $|t ><t|$ in the FHS. Note that N is identified with the integer $(t + 1)$. As emphasized by Pegg and Barnett [32,33], when we deal with phase states, the traceless relation (8.84) is to be used in place of $[b, b^+] = 1$.

The truncated relation (8.77) looks deceptively similar to the generalized quantum condition (5.70) considered in Chapter 5. To avoid confusion of notations, let us rewrite the latter as

$$[\tilde{b}, \tilde{b}^+] = 1 + 2\nu L \tag{8.85}$$

where $\nu \in \mathbb{R}$ and L is an idempotent operator ($L^2 = 1$) that commutes with $\tilde{b}$ and $\tilde{b}^+$. We now remark that (8.85) cannot be transformed to the form (8.77) and hence is not a representative of a truncated scheme. Indeed if we deform (8.77) in terms of the parameters $\lambda, \mu \in \mathbb{R}$ as

$$[B, B^+] = \lambda - \mu N K \tag{8.86}$$

and compare with (8.85), we find for the representations (8.79), and choosing $L = \sigma_3$, the solutions for λ and μ turn out to be

$$\mu = \lambda - 1 = 2\nu \tag{8.87}$$

So λ cannot be put equal to unity since μ and ν will simultaneously vanish.

The root cause of this difficulty is related to the fact that the truncated oscillator is distinct from normal harmonic oscillator possessing infinite dimensional representations [30]. While a generalized quantum condition such as (8.85) speaks for parabosonic oscillators, there are remarkable similarities between the rules obeyed by B, B^+ and those of the parafermionic operators c, c^+ which will be considered now.

(b) Construction of a PSUSY model

Let us consider a truncation at the $(p+1)$th level ($p > 0$, an integer). Then B, B^+ are represented by $(p+1) \times (p+1)$ matrices and (8.77) acquires the form

$$[B, B^+] = 1 - (p+1)K \tag{8.88}$$

where 1 stands for a $(p+1) \times (p+1)$ unit matrix. The irreducible representations of (8.88) are the same as those for the scheme [37] described by a set of operators d and d^+

$$\begin{aligned} [d, d^+d] &= d \\ d^{p+1} &= 0 \\ d^j &\neq 0, \quad j < p+1 \end{aligned} \tag{8.89}$$

Using the decompositions of B and B given in (8.81) we find

$$\begin{aligned} (B)^r &= \sum_{r=0}^{p} \sqrt{p(p-1)(p-2)\ldots(p-r+1)} \\ &\quad |k-r><k| \\ (B^+)^r &= \sum_{r=0}^{p} \sqrt{p(p-1)(p-2)\ldots(p-r+1)} \\ &\quad |k><k-r| \end{aligned} \tag{8.90}$$

Thus

$$(B)^r = (B^+)^r = 0 \quad \text{for } r > p \tag{8.91}$$

Further, the following nontrivial multilinear relation between B and B^+ hold [31,39]

$$B^p B^+ + B^{p-1} B^+ B + \ldots + B^+ B^p = \frac{p(p+1)}{2} B^{p-1} \tag{8.92}$$

along with its hermitean conjugated expression. Of course, these coincide for $p = 1$ and we have the familiar fermionic condition $BB^+ + B^+B = 1$. For $p = 2$ we find from (8.91) and (8.92) the following set

$$(B)^3 = 0 = (B^+)^3 \tag{8.93a}$$

$$B^2 B^+ + BB^+B + B^+B^2 = 3B \tag{8.93b}$$

$$(B^+)^2 B + B^+ B B^+ + B(B^+)^2 = 3B^+ \qquad (8.93c)$$

We notice that, except for the coefficients of B and B^+ in the right-hand-sides of $(8.93b)$ and $(8.93c)$, the above equations resemble remarkabley the trilinear relations which the parafermionic operators c and c^+ obey, namely [25,26]

$$
\begin{aligned}
c^3 &= 0 = (c^+)^3 \\
c^2 c^+ + c c^+ c + c^+ c^2 &= 4c \\
(c^+)^2 c + c^+ c c^+ + c(c^+)^2 &= 4c^+
\end{aligned}
\qquad (8.94)
$$

The generalization of (8.93) to $p = 3$ and higher values are straight-forward.

We may thus interpret B and B^+ to be the annihilation and creation operators of exotic para Fermi oscillators $(p \geq 2)$ governed by the algebraic relation (8.90) - (8.92). Certainly such states here have finite-dimensional representations.

One is therefore motivated to construct a kind of PSUSY scheme of order p in which there will be a symmetry between normal bosons and truncated bosons of order p. So we define a new PSUSY Hamiltonian H_p [which is distinct from either the Rubakov-Spiridonov or the Beckers-Debergh type] generated by parasupercharges Q and Q^+ defined by

$$
\begin{aligned}
(Q)_{\alpha\beta} &= b \otimes B^+ = \sqrt{\beta}\, A_\beta^+ \delta_{\alpha,\beta+1} \\
(Q^+)_{\alpha\beta} &= b^+ \otimes B = \sqrt{\alpha}\, A_\alpha^- \delta_{\alpha+1,\beta}
\end{aligned}
\qquad (8.95)
$$

where $A_\alpha^\pm$ have been defined by (8.55) and (8.56).

One then finds the underlying Hamiltonian H_p to be

$$(H)_{\alpha\beta} = H_\alpha \delta_{\alpha\beta}$$

where

$$
\begin{aligned}
H_r &= -\frac{1}{2}\frac{d^2}{dx^2} + \frac{1}{2}\left(W_r^2 + W_r'\right) \\
&\quad + \frac{1}{2}c_r, \quad r = 1, 2, \ldots, p
\end{aligned}
\qquad (8.96)
$$

$$
H_{p+1} = -\frac{1}{2}\frac{d^2}{dx^2} + \frac{1}{2}\left(W_p^2 - W_p'\right) + \frac{1}{2}c_p
\qquad (8.97)
$$

with the constraint

$$W_{s-1}^2 - W_{s-1}' + c_{s-1} = W_s^2 + W_s' + c_s, \quad s = 2, 3, \ldots p \qquad (8.98)$$

and the parameters $c_1, c_2, \ldots, c_p$ (which have the dimension of energy) obeying

$$c_1 + 2c_2 + \ldots + pc_p = 0 \qquad (8.99)$$

Note that (8.99) is of a different character as compared to (8.61). However, the Hamiltonian and the relationships between the superpotentials as given by (8.96)-(8.98) are similar to the case described in (8.58)-(8.60). As a result the consequences from the two different schemes of PSUSY of order p are identical. To summarize, we have for both the cases the following features to hold

(i) The spectrum is not necessarily positive semidefinite which is the case with SUSYQM.

(ii) The spectrum is $(p+1)$ full degenerate at least above the first p levels while the ground state could be $1, 2, \ldots, p$ fold degenerate depending upon the form of superpotentials.

(iii) One can associate p ordinary SUSYQM Hamiltonians. This is easily checked by writing in place of (8.24) the combination $Q = \sum_{j=1}^{p} \sqrt{j} Q_j$. Q_j's then turn out to be supercharges with $Q_j^2 = 0$.

Finally, we answer the question as to why two seemingly different PSUSY schemes have the same consequences. The answer is that in the case of PSUSY of order p, one has p independent PSUSY charges. In the two schemes of order p that we have addressed, we have merely used two of the p independent forms of Q. It thus transpires that one can very well construct p different PSUSY schemes of order p but all of them will yield almost identical consequences.

8.5 Multidimensional Parasuperalgebras

In this section we explore the PSUSY algebra of order $p = 2$ to study a noninteracting three-level system [40,41] and two bosonic modes possessing different frequencies. Such a system is described by the Hamiltonian

$$H = \frac{1}{2} \sum_{k=1}^{2} \omega_k \left\{ b_k, b_k^+ \right\} + \frac{1}{2} V \tag{8.100}$$

where b_k and b_k^+ are bosonic annihilation and creation operators of the type (2.2)

$$
\begin{aligned}
b_1 &= \frac{1}{\sqrt{2\omega_1}} \left(\frac{d}{dx} + \omega_1 x \right) \\
b_1^+ &= \frac{1}{\sqrt{2\omega_1}} \left(-\frac{d}{dx} + \omega_1 x \right) \\
b_2 &= \frac{1}{\sqrt{2\omega_2}} \left(\frac{d}{dx} + \omega_2 x \right) \\
b_2^+ &= \frac{1}{\sqrt{2\omega_2}} \left(-\frac{d}{dx} + \omega_2 x \right)
\end{aligned} \tag{8.101}
$$

In the above, the frequencies ω_k are distinct ($\omega_1 \neq \omega_2$) and V has a diagonal form to be specified shortly.

In a three-level system there are three possible schemes of configurations of levels, namely the Ξ type, V type, and $\wedge$ type. Note that a three-level atom can sense correlations between electromagnetic field modes with which it interacts [42].

Let us consider the case when the frequencies ω_1 and ω_2 of the bosonic modes are equal to the splitting between various energy levels. Accordingly, we have the following possibilities [40]

$$
\begin{aligned}
\Xi \text{ type}: &\quad \omega_1 = E_1 - E_2, \ \omega_2 = E_2 - E_3 \\
V \text{ type}: &\quad \omega_1 = E_1 - E_2, \ \omega_2 = E_3 - E_2 \\
\wedge \text{ type}: &\quad \omega_1 = E_2 - E_1, \ \omega_2 = E_2 - E_3
\end{aligned} \tag{8.102}
$$

For the Ξ type the Hamiltonian may be considered in terms of the bosonic operators, namely,

$$H_\Xi = \frac{1}{2} \sum_{j=1}^{2} \omega_j \left\{ b_j, b_j^+ \right\} + \text{diag}(E_1, E_2, E_3) \tag{8.103}$$

The transition operators between levels 1 and 2 which are denoted as $t_1^\pm$ and those between levels 2 and 3 which are denoted as $t_2^\pm$ are explicitly given by

$$t_1^+ = \frac{1}{\sqrt{2}} \begin{pmatrix} 0 & x & 0 \\ 0 & 0 & 0 \\ 0 & 0 & 0 \end{pmatrix}$$

$$t_2^+ = \frac{1}{\sqrt{2}} \begin{pmatrix} 0 & 0 & 0 \\ 0 & 0 & y \\ 0 & 0 & 0 \end{pmatrix} \tag{8.104}$$

along with their conjugated representations. In (8.104), x and y are nonzero real quantities.

The above forms for t_1^+ and t_2^+ induce charges Q_1^+ and Q_2^+ given by

$$Q_1^+ = \frac{1}{\sqrt{\omega_1}} \begin{pmatrix} 0 & b_1 & 0 \\ 0 & 0 & 0 \\ 0 & 0 & 0 \end{pmatrix}$$

$$Q_2^+ = \frac{1}{\sqrt{\omega_2}} \begin{pmatrix} 0 & 0 & 0 \\ 0 & 0 & b_2 \\ 0 & 0 & 0 \end{pmatrix} \tag{8.105}$$

If H_Ξ is redefined slightly to have a change in the zero-point energy so that

$$H_\Xi = \frac{1}{2} \sum_{j=1}^{2} \omega_j \left\{ b_j, b_j^+ \right\}$$
$$+ \operatorname{diag}(E_1 - E_3, 2E_2 - E_1 - E_3, E_3 - E_1) \tag{8.106}$$

where the energy-frequency relationships are provided by (8.102), it follows that H_Ξ along with $Q_1^\pm$ and $Q_2^\pm$ obey the relations

$$(Q_i^\pm)^2 = 0 \quad i = 1, 2$$
$$\left[H, Q_i^\pm\right] = 0 \quad i = 1, 2$$
$$Q_1^+ \left(Q_1^- Q_1^+ + Q_2^+ Q_2^-\right) = Q_1^+ H$$
$$\left(Q_1^- Q_1^+ + Q_2^+ Q_2^-\right) Q_2^+ = Q_2^+ H \tag{8.107}$$

along with their hermitean-conjugated counterparts.

We are thus led to a generalized scheme in which superpotentials are introduced, in place of bosonic operators in (8.105). Thus we define in two dimensions

$$Q_1^+ = \frac{1}{\sqrt{2}} \begin{pmatrix} 0 & A_1^+(x) & 0 \\ 0 & 0 & 0 \\ 0 & 0 & 0 \end{pmatrix}$$

$$Q_2^+ = \frac{1}{\sqrt{2}} \begin{pmatrix} 0 & 0 & 0 \\ 0 & 0 & A_2^+(y) \\ 0 & 0 & 0 \end{pmatrix} \tag{8.108}$$

and read off form (8.107)

$$H = \text{diag}(H_1, H_2, H_3) \tag{8.109}$$

where

$$\begin{aligned}
H_1 &= \frac{1}{2}\left[A_1^+(x)A_1^-(x) + A_2^+(y)A_2^-(y)\right] \\
&= \frac{1}{2}\left[-\frac{d^2}{dx^2} - \frac{d^2}{dy^2} + W_1^2(x) + W_2^2(y)\right. \\
&\qquad \left. +W_1'(x) + W_2'(y)\right] \tag{8.110}
\end{aligned}$$

$$\begin{aligned}
H_2 &= \frac{1}{2}\left[A_1^-(x)A_1^+(x) + A_2^+(y)A_2^-(y)\right] \\
&= \frac{1}{2}\left[-\frac{d^2}{dx^2} - \frac{d^2}{dy^2} + W_1^2(x) + W_2^2(y)\right. \\
&\qquad \left. -W_1'(x) + W_2'(y)\right] \tag{8.111}
\end{aligned}$$

$$\begin{aligned}
H_3 &= \frac{1}{2}\left[A_1^-(x)A_1^+(x) + A_2^-(y)A_2^+(y)\right] \\
&= \frac{1}{2}\left[-\frac{d^2}{dx^2} - \frac{d^2}{dy^2} + W_1^2(x) + W_2^2(y)\right. \\
&\qquad \left. -W_1'(x) - W_2'(y)\right] \tag{8.112}
\end{aligned}$$

To derive H_1, H_2, H_3 a constraint like (8.16) was not needed. These components of the Hamiltonian H together with Q_1^+ and Q_2^+ defined by (8.108) offer a two-dimensional generalization of the conventional PSUSY schemes. Note that the underlying algebra is provided by (8.107).

It is pointless to mention that (8.110)-(8.112) are consistent with (8.106) when $W_1(x) = \omega_1 x$ and $W_2(y) = \omega_2 y$. We should also point out that if we define Q and Q^+ according to (8.24) using (8.108), then while Q and Q^+ obey $Q^3 = (Q^+)^3 = 0$, Q_1^+ and Q_2^+ play the role of supercharges.

This concludes our discussion on the generalized parasuperalgebras associated with the three-level system of Ξ type. The transition operators, and consequently the supercharges for the systems V and $\wedge$ can be similarly built, and generalized schemes such as the one described for the Ξ type can be set up.

8.6 References

[1] V.A. Rubakov and V.P. Spiridonov, *Mod. Phys. Lett.*, **A3**, 1337, 1988.

[2] J. Beckers and N. Debergh, *Nucl. Phys.*, **B340**, 770, 1990.

[3] S. Durand and L. Vinet, *Phys. Lett.*, **A146**, 299, 1990.

[4] J. Beckers and N. Debergh, *Mod. Phys. Lett.*, **A4**, 2289, 1989.

[5] J. Beckers and N. Debergh, *J. Math. Phys.*, **31**, 1523, 1990.

[6] J. Beckers and N. Debergh, *J. Phys. A: Math. Gen.*, **23**, L751, 1990.

[7] J. Beckers and N. Debergh, *J. Phys. A: Math. Gen.*, **23**, L1073, 1990.

[8] J. Beckers and N. Debergh, *Z. Phys.*, **C51**, 519, 1991.

[9] S. Durand and L. Vinet, *J. Phys. A: Math. Gen.*, **23**, 3661, 1990.

[10] S. Durand, R. Floreanimi, M. Mayrand, and L. Vinet, *Phys. Lett.*, **B233**, 158, 1989.

[11] V. Spiridonov, *J. Phys. A: Math. Gen.*, **24**, L529, 1991.

[12] A.A. Andrianov and M.V. Ioffe, *Phys. Lett.*, **B255**, 543, 1989.

[13] A.A. Andrianov, M.V. Ioffe, V. Spiridonov, and L. Vinet, *Phys. Lett.*, **B272**, 297, 1991.

[14] V. Merkel, *Mod. Phys. Lett.*, **A5**, 2555, 1990.

[15] V. Merkel, *Mod. Phys. Lett.*, **A6**, 199, 1991.

[16] S. Durand, M. Mayrand, V. Spiridonov, and L. Vinet, *Mod. Phys. Lett.*, **A6**, 3163, 1991.

[17] H.S. Green, *Phys. Rev.*, **90**, 270, 1953.

[18] D.V. Volkov, *Zh. Eksp. Teor. Fiz.*, **38**, 519, 1960.

[19] D.V. Volkov, *Zh. Eksp. Teor. Fiz.*, **39**, 1560, 1960.

[20] O.W. Greenberg and A.M.L. Messiah, *Phys. Rev.*, **B138**, 1155, 1965.

[21] Y. Ohnuki and S. Kamefuchi, *Quantum Field Theory and Parastatistics*, University of Tokyo, Tokyo, 1982.

[22] A.A. Andrianov, M.V. Ioffe, and D. Nishnianidze, *Theor. Math. Phys.*, **104**, 1129, 1995.

[23] A.A. Andrianov, M.V. Ioffe, and V. Spiridonov, *Phys. Lett.*, **A174**, 273, 1993.

[24] A. A. Andrianov, F. Cannata, J.P. Dedonder, and M.V. Ioffe, *Int. J. Mod. Phys.*, **A10**, 2683, 1995.

[25] A. Khare, *J. Math. Phys.*, **34**, 1277, 1993.

[26] A. Khare, *J. Phys. A: Math. Gen.*, **25**, l749, 1992.

[27] M. Daoud and V. Hassouni, *Int. J. Theor. Phys.*, **37**, 2021, 1998.

[28] C. Quesne and N. Vansteenkiste, *Phys. Lett.*, **A240**, 21, 1998.

[29] C. Quesne and N. Vansteenkiste, C_λ-*extended Oscillator Algebras and Some of their Deformations and Applications to Quantum Mechanics*, preprint, 1999.

[30] H.A. Buchdahl, *Am. J. Phys.*, **35**, 210, 1967.

[31] B. Bagchi and P.K. Roy, *Phys. Lett.*, **A200**, 411, 1995.

[32] D.T. Pegg and S.M. Barnett, *Europhys. Lett.*, **6**, 483, 1988.

[33] D.T. Pegg and S.M. Barnett, *Phys. Rev.*, **A39**, 1665, 1989.

[34] X. Ma and W. Rhodes, *Phys. Rev.*, **A43**, 2576, 1991.

[35] J-S. Peng and G-X Li, *Phys. Rev.*, **A45**, 3289, 1992.

[36] A.D. Wilson - Gordon, V. Buzeck, and P.L. Knight, *Phys. Rev.*, **A44**, 7647, 1991.

[37] R. Kleeman, *J. Aust. Math. Soc.*, **B23**, 52, 1981.

[38] L-M Kuang, *J. Phys. A: Math. Gen.*, **26**, L1079, 1993.

[39] B. Bagchi, S.N. Biswas, A. Khare, and P.K. Roy, *Pramana J. of Phys.*, **49**, 199, 1997.

[40] V.V. Semenov, *J. Phys. A: Math. Gen.*, **25**, L511, 1992.

[41] B. Bagchi and K. Samanta, *Multidimensional Parasuperalgebras and a Noninteracting N-level System*, preprint, 1992.

[42] V. Buzeck, P.L. Knight, and I.K. Kudryartsev, *Phys. Rev.*, **A44**, 1931, 1991.

Appendix A

The D-dimensional Schroedinger Equation in a Spherically Symmetric Potential $V(r)$

In Cartesian coordinates, the Schroedinger equation under the influence of a potential $V(r)$ reads

$$-\frac{\hbar^2}{2m}\nabla_D^2\psi(\overrightarrow{r}) + V(r)\psi(\overrightarrow{r}) = E\psi(\overrightarrow{r}) \qquad (A1)$$

where

$$\overrightarrow{r} = (x_1, x_2, \ldots x_D), \; r = |\overrightarrow{r}| \qquad (A2)$$

$$\nabla_D^2 = \frac{\partial}{\partial x_i}\frac{\partial}{\partial x_i}$$

Our task is to transform $(A1)$ to D-dimensional polar coordinates. The latter are related to the Cartesian coordinates by

$$
\left.
\begin{aligned}
x_1 &= r\cos\theta_1\sin\theta_2\sin\theta_3\ldots\ldots\sin\theta_{D-1} \\[2mm]
x_2 &= r\sin\theta_1\sin\theta_2\sin\theta_3\ldots\ldots\sin\theta_{D-1} \\[2mm]
x_3 &= r\cos\theta_2\sin\theta_3\sin\theta_4\ldots\ldots\sin\theta_{D-1} \\[2mm]
x_4 &= r\cos\theta_3\sin\theta_4\sin\theta_5\ldots\ldots\sin\theta_{D-1} \\[1mm]
&\;\vdots \\[1mm]
x_j &= r\cos\theta_{j-1}\sin\theta_j\sin\theta_{j+1}\ldots\ldots\sin\theta_{D-1} \\[1mm]
&\;\vdots \\[1mm]
x_{D-1} &= r\cos\theta_{D-1}\sin\theta_{D-1} \\
x_D &= r\cos\theta_{D-1}
\end{aligned}
\right\} \qquad (A3)
$$

where
$$
\begin{aligned}
D &= 3, 4, 5 \ldots \\
0 &< r < \infty \\
0 &\leq \theta_1 < 2\pi \\
0 &\leq \theta_j \leq \pi, \ j = 2, 3, \ldots D - 1
\end{aligned}
\tag{A4}
$$

The Laplacian ∇_D^2 can be written as

$$
\nabla_D^2 = \frac{1}{h} \sum_{i=0}^{D-1} \frac{\partial}{\partial \theta_i} \left(\frac{h}{h_i^2} \frac{\partial}{\partial \theta_i} \right)
\tag{A5}
$$

where

$$
\theta_0 = r, \ h = \prod_{i=0}^{D-1} h_i
\tag{A6}
$$

and the scale factors h_i are given by

$$
h_i^2 = \sum_{k=1}^{D} \left(\frac{\partial x_k}{\partial \theta_i} \right)^2 \ i = 0, 1, 2, \ldots, D - 1
\tag{A7}
$$

Explicitly

$$
\begin{aligned}
h_0^2 &= \left(\frac{\partial x_1}{\partial \theta_0} \right)^2 + \left(\frac{\partial x_2}{\partial \theta_0} \right)^2 + \ldots + \left(\frac{\partial x_D}{\partial \theta_0} \right)^2 = 1 \\[2mm]
h_1^2 &= \left(\frac{\partial x_1}{\partial \theta_1} \right)^2 + \left(\frac{\partial x_2}{\partial \theta_1} \right)^2 = r^2 \sin^2 \theta_2 \sin^2 \theta_3 \ldots \sin^2 \phi_{D-1} \\[2mm]
h_2^2 &= \left(\frac{\partial x_1}{\partial \theta_2} \right)^2 + \left(\frac{\partial x_2}{\partial \theta_2} \right)^2 + \left(\frac{\partial x_3}{\partial \theta_2} \right)^2 \\[2mm]
&= r^2 \sin^2 \theta_3 \sin^2 \theta_4 \ldots \sin^2 \phi_{D-1} \\[1mm]
&\vdots \\
h_j^2 &= r^2 \sin^2 \theta_{j+1} \sin^2 \theta_{j+2} \ldots \sin^2 \theta_{D-1} \\[1mm]
&\vdots \\
h_{D-1}^2 &= r^2
\end{aligned}
\tag{A8}
$$

Thus h is

$$
\begin{aligned}
h &= h_0 h_1 \ldots h_{D-1} \\[2mm]
&= r^{D-1} \sin \theta_2 \sin^2 \theta_3 \sin^3 \theta_4 \ldots \sin^{D-2} \theta_{D-1}
\end{aligned}
\tag{A9}
$$

From ($A5$), the first term of ∇_D^2 is

$$
= \frac{1}{h}\frac{\partial}{\partial\theta_0}\frac{h}{h_0^2}\frac{\partial}{\partial\theta_0}
$$

$$
= \frac{1}{r^{D-1}\sin\theta_2\sin^2\theta_3\dots\sin^{D-2}\theta_{D-1}}\frac{\partial}{\partial r}r^{D-1}\sin\theta_2\dots\sin^{D-2}\theta_{D-1}\frac{\partial}{\partial r}
$$

$$
= \frac{1}{r^{D-1}}\frac{\partial}{\partial r}r^{D-1}\frac{\partial}{\partial r}.
$$

$$(A10)$$

The last term of ∇_D^2 is

$$
= \frac{1}{h}\frac{\partial}{\partial\theta_{D-1}}\frac{h}{h_{D-1}^2}\frac{\partial}{\partial\theta_{D-1}}
$$

$$
= \frac{1}{r^{D-1}\sin\theta_2\sin^2\theta_3\dots\sin^{D-2}\theta_{D-1}}\frac{\partial}{\partial\theta_{D-1}}
$$

$$
\frac{r^{D-1}\sin\theta_2\dots\sin^{D-2}\theta_{D-1}}{r^2}\frac{\partial}{\partial\theta_{D-1}}
$$

$$
= \frac{1}{r^2\sin^{D-2}\theta_{D-1}}\frac{\partial}{\partial\theta_{D-1}}\sin^{D-2}\theta_{D-1}\frac{\partial}{\partial\theta_{D-1}}
$$

$$(A11)$$

Other terms of ∇_D^2 are of the forms

$$
= \frac{1}{h}\frac{\partial}{\partial\theta_j}\frac{h}{h_j^2}\frac{\partial}{\partial\theta_j}
$$

$$
= \frac{1}{r^{D-1}\sin\theta_2\dots\sin^{j-1}\theta_j\sin^j\theta_{j+1}\dots\sin^{D-2}\theta_{D-1}}\frac{\partial}{\partial\theta_j}
$$

$$
\frac{r^{D-1}\sin\theta_2\dots\sin^{j-1}\theta_j\dots\sin^{D-2}\theta_{D-1}}{r^2\sin^2\theta_{j+1}\dots\sin^2\theta_{D-1}}\frac{\partial}{\partial\theta_j}
$$

$$
= \frac{1}{r^2\sin^2\theta_{j+1}\dots\sin^2\theta_{D-1}}\left(\frac{1}{\sin^{j-1}\theta_j}\frac{\partial}{\partial\theta_j}\sin^{j-1}\theta_j\frac{\partial}{\partial\theta_j}\right)
$$

$$(A12)$$

Using $(A10)$-$(A12)$, we get from $(A5)$ the representation

$$\nabla_D^2 = \frac{1}{r^{D-1}}\frac{\partial}{\partial r}r^{D-1}\frac{\partial}{\partial r} + \frac{1}{r^2}\sum_{j=1}^{D-2}\frac{1}{\sin^2\theta_{j+1}\ldots\sin^2\theta_{D-1}}$$

$$\left(\frac{1}{\sin^{j-1}\theta_j}\frac{\partial}{\partial\theta_j}\sin^{j-1}\theta_j\frac{\partial}{\partial\theta_j}\right) \tag{A13}$$

$$+\frac{1}{r^2}\left(\frac{1}{\sin^{D-2}\theta_{D-1}}\frac{\partial}{\partial\theta_{D-1}}\sin^{D-2}\theta_{D-1}\frac{\partial}{\partial\theta_{D-1}}\right)$$

We note also that the Laplacian ∇_D^2 obeys the relation

$$\nabla_D^2 = \frac{1}{r^{D-1}}\frac{\partial}{\partial r}r^{D-1}\frac{\partial}{\partial r} - \frac{L_{D-1}^2}{r^2} \tag{A14}$$

with

$$L_n^2 = \sum_{i,j}L_{ij}L_{ij}, \ i = 1, 2, \ldots j - 1 \tag{A15}$$

$$j = 2, \ldots D$$

and the angular momentum components L_{ij} are defined as the skew symmetric tensors

$$L_{ij} = -L_{ji}$$
$$= x_i p_j - x_j p_i, \ i = 1, 2, \ldots j - 1 \tag{A16}$$
$$j = 2, \ldots D$$

To prove $(A14)$, we first note that we can express p_k as

$$p_k = -i\hbar\frac{\partial}{\partial x_k} = -i\hbar\sum_{r=0}^{D-1}\left(\frac{\partial\theta_r}{\partial x_k}\right)\frac{\partial}{\partial\theta_r}$$

$$= -i\hbar\sum_{r=0}^{D-1}\left(\frac{1}{h_r^2}\frac{\partial x_k}{\partial\theta_r}\right)\frac{\partial}{\partial\theta_r} \tag{A17}$$

where we have used the relations

$$\sum_{l=0}^{D-1}\frac{\partial x_l}{\partial x_i}\frac{\partial x_l}{\partial\theta_r} = \delta_{ir}\ h_i^2,$$

$$\sum_{l=0}^{D-1}\frac{\partial\theta_i}{\partial x_k}\frac{\partial x_l}{\partial\theta_i} = \delta_{kl} \tag{A18}$$

We next note that the following commutation relation holds (see Appendix B)

$$[L_{ij}, L_{kl}] = i\hbar\delta_{jl}L_{ik} + i\hbar\delta_{ik}L_{jl} - i\hbar\delta_{jk}L_{il} - i\hbar\delta_{il}L_{jk} \qquad (A19)$$

Further if we set

$$L_k^2 = \sum_{i,j} L_{ij}L_{ij}, \quad i = 1, 2, \ldots j - 1$$
$$j = 2, 3, \ldots k + 1 \qquad (A20)$$

we can obtain [see Appendix B]

$$L_1^2 = -\frac{\partial^2}{\partial\theta_1^2}$$

$$L_2^2 = -\left(\frac{1}{\sin\theta_2}\frac{\partial}{\partial\theta_2}\sin\theta_2\frac{\partial}{\partial\theta_2} - \frac{L_1^2}{\sin^2\theta_2}\right)$$

$$\vdots$$

$$L_k^2 = -\left[\frac{1}{\sin^{k-1}\theta_k}\frac{\partial}{\partial\theta_k}\sin^{k-1}\theta_k\frac{\partial}{\partial\theta_k} - \frac{L_{k-1}^2}{\sin^2\theta_k}\right]$$

$$\vdots$$

$$L_{D-1}^2 = \left(\frac{1}{\sin^{D-2}\theta_{D-1}}\frac{\partial}{\partial\theta_{D-1}}\sin^{D-2}\theta_{D-1}\frac{\partial}{\partial\theta_{D-1}} - \frac{L_{D-2}^2}{\sin^2\theta_{D-1}}\right)$$
$$(A21)$$

Therefore, from $(A13)$, we get

$$\nabla_D^2 = \frac{1}{r^{D-1}}\frac{\partial}{\partial r}r^{D-1}\frac{\partial}{\partial r} - \frac{L_{D-1}^2}{r^2} \qquad (A22)$$

From $(A21)$ it is clear that since $\theta_1, \theta_2, \ldots \theta_{D-1}$ are independent, the operators $L_1^2, L_2^2 \ldots L_{D-1}^2$ mutually commute. Hence, they have a common eigenfunction $Y(\theta_1, \theta_2, \ldots, \theta_{D-1})$. Let us write

$$L_k^2 Y(\theta_1, \theta_2, \ldots, \theta_{D-1}) = \lambda_k Y(\theta_1, \theta_2 \ldots \theta_{D-1}) \qquad (A23)$$

where λ_k is the eigenvalue of L_k^2. Since the potential function is independent of t, $Y(\theta_1, \theta_2, \ldots, \theta_{D-1})$ can be expressed as

$$Y(\theta_1, \theta_2, \ldots, \theta_{D-1}) = \prod_{k=1}^{D-1} \Theta_k(\theta_k) \qquad (A24)$$

Then we get from $(A23)$

$$\frac{Y}{\Theta_1(\theta_1)}L_1^2\Theta_1(\theta_1) = \lambda_1 Y \qquad (A25)$$

where we note that L_1^2 is dependent on θ_1 only. In other words

$$L_1^2\Theta_1(\theta_1) = \lambda_1\Theta_1(\theta_1) \qquad (A26)$$

Similarly from $L_2^2 Y = \lambda_2 Y$ we get

$$L_2^2\Theta_1(\theta_1)\Theta_2(\theta_2) = \lambda_2\Theta_1(\theta_1)\Theta_2(\theta_2) \qquad (A27)$$

where L_2^2 is dependent on θ_1 and θ_2 only.
Using the explicit form of L_2^2 we have

$$-\left[\frac{L_1^2}{\sin^2\theta_2} - \frac{1}{\sin\theta_2}\frac{\partial}{\partial\theta_2}\left(\sin\theta_2\frac{\partial}{\partial\theta_2}\right)\right]\Theta_1(\theta_1)\Theta_2(\theta_2)$$

$$= \lambda_2\Theta_1(\theta_1)\Theta_2(\theta_2) \qquad (A28)$$

or

$$\frac{\Theta_2(\theta_2)}{\sin^2\theta_2}L_1^2\Theta_1(\theta_1) - \frac{\Theta_1(\theta_1)}{\sin\theta_2}\frac{\partial}{\partial\theta_2}\sin\theta_2\frac{\partial}{\partial\theta_2}\Theta_2(\theta_2)$$

$$= \lambda_2\Theta_1(\theta_1)\Theta_2(\theta_2) \qquad (A29)$$

or

$$\frac{\Theta_2(\theta_2)}{\sin^2\theta_2}\lambda_1\Theta_1(\theta_1) - \frac{\Theta_1(\theta_1)}{\sin\theta_2}\frac{\partial}{\partial\theta_2}\sin\theta_2\frac{\partial}{\partial\theta_2}\Theta_2(\theta_2)$$

$$= \lambda_2\Theta_1(\theta_1)\Theta_2(\theta_2) \qquad (A30)$$

or

$$\frac{\lambda_1\Theta_2(\theta_2)}{\sin^2\theta_2} - \frac{\partial^2}{\partial\theta_2^2}\Theta_2(\theta_2) - \cot\theta_2\frac{\partial}{\partial\theta_2}\Theta_2(\theta_2) = \lambda_2\Theta_2(\theta_2) \qquad (A31)$$

implying

$$-\left(\frac{\partial^2}{\partial\theta_2^2} + \cot\theta_2\frac{\partial}{\partial\theta_2} - \frac{\lambda_1}{\sin^2\theta_2}\right)\Theta_2(\theta_2) = \lambda_2\Theta_2(\theta_2) \qquad (A32)$$

Let us suppose that

$$-\left(\frac{\partial^2}{\partial\theta_k^2} + (k-1)\cot\theta_k\frac{\partial}{\partial\theta_k} - \frac{\lambda_{k-1}}{\sin^2\theta_k}\right)\Theta_k(\theta_k) = \lambda_k\Theta_k(\theta_k) \qquad (A33)$$

Looking at the eigenvalue equation

$$L_{k+1}^2 Y = \lambda_{k+1} Y \tag{A34}$$

which on expansion becomes

$$-\left(\frac{1}{\sin^k \theta_{k+1}} \frac{\partial}{\partial \theta_{k+1}} \sin^k \theta_{k+1} \frac{\partial}{\partial \theta_{k+1}} - \frac{L_k^2}{\sin^2 \theta_{k+1}} \right) Y = \lambda_{k+1} Y \tag{A35}$$

can also be expressed as

$$-\frac{Y}{\Theta_{k+1}(\theta_{k+1})} \frac{1}{\sin^k \theta_{k+1}} \frac{\partial}{\partial \theta_{k+1}} \sin^k \theta_{k+1} \frac{\partial}{\partial \theta_{k+1}} \Theta_{k+1}(\theta_{k+1})$$

$$+ \frac{\lambda_k Y}{\sin^2 \theta_{k+1}} = \lambda_{k+1} Y$$

or

$$-\left(\frac{\partial^2}{\partial \theta_{k+1}^2} + k \cot \theta_{k+1} \frac{\partial}{\partial \theta_{k+1}} - \frac{\lambda_k}{\sin^2 \theta_{k+1}} \right) \Theta_{k+1}(\theta_{k+1})$$

$$= \lambda_{k+1} \Theta_{k+1}(\theta_{k+1}) \tag{A36}$$

The above shows that if $(A33)$ holds for k, it also holds for $k+1$ as well. Now, we have clearly shown in $(A32)$, that it holds for $k = 2$. Therefore, by the principle of induction, $(A33)$ holds $\forall k = 2, 3, \ldots D-1$.

Let us write

$$L_k^2(\lambda_{k-1}) = -\left(\frac{\partial^2}{\partial \theta_k^2} + (k-1) \cot \theta_k \frac{\partial}{\partial \theta_k} - \frac{\lambda_{k-1}}{\sin^2 \theta_k} \right) \tag{A37}$$

Hence we have

$$L_1^2 \Theta_1(\theta_1) = \lambda_1 \Theta_1(\theta_1) \tag{A38}$$

$$L_k^2(\lambda_{k-1}) \Theta_k(\theta_k) = \lambda_k \Theta_k(\theta_k), \quad k = 2, 3, \ldots, D-1 \tag{A39}$$

We now turn to the eigenvalues λ_k. For $k = 2$

$$\lambda_1 = l_1^2 \tag{A40}$$

$$\lambda_2 = l_2(l_2 + 1) \tag{A41}$$

where

$$l_2 = 0, 1, 2, \ldots \tag{A42}$$

$$l_1 = -l_2, -l_2 + 1, \ldots l_2 - 1, l_2. \tag{A43}$$

Let us assume

$$\lambda_{k-1} = l_{k-1}(l_{k-1} + k - 2) \tag{A44}$$

where l_{k-1} is an integer. Setting

$$L_k^+(l_{k-1}) = \frac{\partial}{\partial \theta_k} - l_{k-1} \cot \theta_k$$

$$L_k^-(l_{k-1}) = -\frac{\partial}{\partial \theta_k} - (l_{k-1} + k - 2) \cot \theta_k \tag{A45}$$

it follows by induction

$$\lambda_k = l_k(l_k + k - 1) \tag{A46}$$

We finally have from $(A24)$ and $(A39)$

$$L_{D-1}^2 Y_{l_{D-1}, l_{D-2}, \ldots, l_2, l_1}(\theta_1, \theta_2, \ldots, \theta_{D-1})$$

$$= l_{D-1}(l_{D-1} + D - 2) Y_{l_{D-1}, l_{D-2}, \ldots, l_2, l_1}(\theta_1, \ldots, \theta_{D-1}) \tag{A47}$$

where $Y_{l_{D-1}, l_{D-2}, \ldots, l_2, l_1}$ are the generalised spherical harmonics and

$$\begin{aligned} l_{D-1} &= 0, 1, 2, \ldots \\ L_{D-2} &= 0, 1, 2, \ldots, l_{D-1}, \\ &\vdots \\ l_1 &= -l_2, -l_2 + 1, \ldots, l_2 - 1, l_2 \end{aligned} \tag{A48}$$

Substituting in $(A1)$

$$\psi(r) = R(r) Y_{l_{D-1}, l_{D-2}, \ldots l_1}(\theta_1, \theta_2, \ldots, \theta_{D-1}) \tag{A49}$$

and using $(A22)$ and $(A47)$, we obtain the radial part of the Schroedinger equation as

$$-\frac{\hbar^2}{2m} \left[\frac{d^2}{dr^2} + \frac{D-1}{r} \frac{d}{dr} + \frac{l(l + D - 2)}{r^2} \right] R(r) + V(r) R(r) = E R(r) \tag{A50}$$

To eliminate the first derivative, we make the substitution

$$R(r) = r^{(1-N)/2}u(r)$$

so that $(A21)$ reduces to

$$-\frac{\hbar^2}{2m}\left[\frac{d^2u}{dr^2} - \frac{\alpha_l}{r^2}u(r)\right] + V(r)u(r) = Eu(r) \qquad (A51)$$

where $\alpha_l = \frac{1}{4}(D-1)(D-3) + l(l+D-2)$.

Appendix B

Derivation of the Results $(A19)$ and $(A21)$

Consider the angular momentum components defined by

$$L_{ij} = -L_{ji} = x_i p_j - x_j p_i, \quad i = 1, 2, \ldots, j-1 \ \ j = 2, \ldots D \quad (B1)$$

We also set

$$L_k^2 = \sum_{i,j} L_{ij} L_{ij} \ i = 1, 2, \ldots j-1 \ \ j = 2, 3, \ldots, k+1 \quad (B2)$$

In this appendix, let us first prove the following angular momentum commutation relation $(\hbar = 1)$

$$[L_{ij}, L_{kl}] = i\delta_{jl} L_{ik} + i\delta_{ik} L_{jl} - i\,\delta_{jk} L_{il} - i\delta_{il} L_{jk} \quad (B3)$$

We make use of the commutation relations

$$[x_i, p_j] = i\delta_{ij} \quad (B4)$$

and

$$[x_i, x_j] = 0 = [p_i, p_j] \quad (B5)$$

to note that the right-hand-side of $(B3)$

$$= (x_j p_l - p_l x_j)(x_i p_k - x_k p_i) + (x_i p_k - p_k x_i)(x_j p_l - x_l p_j)$$

$$-(x_j p_k - p_k x_j)(x_i p_l - x_l p_i) - (x_i p_l - p_l x_i)(x_j p_k - x_k p_j)$$

$$= x_j p_l x_i p_k - p_l x_j x_i p_k - x_j p_l x_k p_i + p_l x_j x_k p_i + x_i p_k x_j p_l$$

$$-p_k x_i x_j p_l - x_i p_k x_l p_j + p_k x_i x_l p_j - x_j p_k x_i p_l + p_k x_j x_i p_l$$

$$+x_j p_k x_l p_i - p_k x_j x_l p_i - x_i p_l x_j p_k + p_l x_i x_j p_k + x_i p_l x_k p_j - p_l x_i x_k p_j$$

$$= x_j p_l (p_k x_i + i\delta_{ik}) + x_i p_k (p_l x_j + i\delta_{jl}) - x_j p_k (p_l x_i + i\delta_{il})$$

$$-x_i p_l (p_k x_j + i\delta_{jk}) - x_j p_l (p_i x_k + i\delta_{ik}) - x_i p_k (p_j x_l + i\delta_{jl})$$

$$+x_j p_k (p_i x_l + i\delta_{il}) + x_i p_l (p_j x_k + i\delta_{jk}) + p_l x_k (x_j p_i - p_j x_i)$$

$$+p_k x_l (x_i p_j - x_j p_i)$$

$$= x_j p_i (-p_l x_k + p_k x_l) - x_i p_j (-p_l x_k + p_k x_l) + (p_k x_l - p_l x_k) L_{ij}$$

[using the definition $(B1)$]

$$= (x_j p_i - x_i p_j)(p_k x_l - p_l x_k) + (p_k x_l - p_l x_k) L_{ij}$$

$$= L_{ji}(x_l p_k - x_k p_l) + (x_l p_k - x_k p_l) L_{ij}$$

$$= L_{ij} L_{kl} - L_{kl} L_{ij}.$$
$$= \text{Left-hand-side of } (B3).$$

We next write

$$L_1^2 f \quad = \quad L_{12}^2 f = L_{12} L_{12} f = (x_1 p_2 - x_2 p_1)(x_1 p_2 - x_2 p_1) f$$

$$= \quad -\left(x_1 \frac{\partial}{\partial x_2} - x_2 \frac{\partial}{\partial x_1}\right)\left(x_1 \frac{\partial}{\partial x_2} - x_2 \frac{\partial_2}{\partial x_1}\right) f \qquad (B6)$$

for an arbitrary function f. Now

$$\left(x_1 \frac{\partial}{\partial x_2} - x_2 \frac{\partial}{\partial x_1} \right) f$$

$$= \sum_{j=0}^{D-1} \frac{1}{h_j^2} \left(x_1 \frac{\partial x_2}{\partial \theta_j} - x_2 \frac{\partial x_1}{\partial \theta_j} \right) \frac{\partial f}{\partial \theta_j}$$

$$= \sum_{j=0}^{D-1} \frac{x_1^2}{h_j^2} \left\{ \frac{\partial}{\partial \theta_j} \left(\frac{x_2}{x_1} \right) \right\} \frac{\partial f}{\partial \theta_j}$$

$$= \sum_{j=0}^{D-1} \frac{x_1^2}{h_j^2} \frac{\partial}{\partial \theta_j} (\tan \theta_1) \frac{\partial f}{\partial \theta_j}$$

$$= \frac{x_1^2}{h_1^2} \sec^2 \theta_1 \frac{\partial f}{\partial \theta_1}$$

$$= \frac{r^2 \cos^2 \theta_1 \sin^2 \theta_2 \ldots \sin^2 \theta_{D-1}}{r^2 \sin^2 \theta_2 \ldots \sin^2 \theta_{D-1}} \sec^2 \theta_1 \frac{\partial f}{\partial \theta_1}$$

$$= \frac{\partial f}{\partial \theta_1} \qquad (B7)$$

Therefore

$$L_1^2 = -\frac{\partial^2}{\partial \theta_1^2} \qquad (B8)$$

Next

$$L_2^2 = \sum_{j=2,3}^{j-1} \sum_{i=1} L_{ij} L_{ij} = L_1^2 + L_{13} L_{13} + L_{23} L_{23} \qquad (B9)$$

where

$$L_{13}^2 f = (x_1 p_3 - x_3 p_1)^2 f = -\left(x_1 \frac{\partial}{\partial x_3} - x_3 \frac{\partial}{\partial x_1} \right)^2 f \qquad (B10)$$

$$L_{23}^2 f = (x_2 p_3 - x_3 p_2)^2 f = -\left(x_2 \frac{\partial}{\partial x_3} - x_3 \frac{\partial}{\partial x_2} \right)^2 f \qquad (B11)$$

Consider

$$\left(x_2\frac{\partial}{\partial x_3} - x_3\frac{\partial}{\partial x_2}\right)f$$

$$= \sum_{j=0}^{D-1}\frac{1}{h_j^2}\left(x_2\frac{\partial x_3}{\partial\theta_j} - x_3\frac{\partial x_2}{\partial\theta_j}\right)\frac{\partial f}{\partial\theta_j}$$

$$= \sum_{j=0}^{D-1}\frac{1}{h_j^2}x_2^2\frac{\partial}{\partial\theta_j}\left(\frac{x_3}{x_2}\right)\frac{\partial f}{\partial\theta_j}$$

$$= \sum_{j=0}^{D-1}\frac{x_2^2}{h_j^2}\frac{\partial}{\partial\theta_j}(\cot\theta_2\mathrm{cosec}\theta_1)\frac{\partial f}{\partial\theta_j}$$

$$= \frac{x_2^2}{h_1^2}\frac{\partial}{\partial\theta_1}(\cot\theta_2\mathrm{cosec}\theta_1)\frac{\partial f}{\partial\theta_1} + \frac{x_2^2}{h_2^2}\frac{\partial}{\partial\theta_2}(\cot\theta_2\mathrm{cosec}\theta_1)\frac{\partial f}{\partial\theta_2}$$

$$= -\frac{r^2\sin^2\theta_1\sin^2\theta_2\ldots\sin^2\theta_{D-1}}{r^2\sin^2\theta_2\ldots\sin^2\theta_{D-1}}\cot\theta_2\mathrm{cosec}\theta_1\cot\theta_1\frac{\partial f}{\partial\theta_1}$$

$$-\frac{r^2\sin^2\theta_1\ldots\sin^2\theta_{D-1}}{r^2\sin^2\theta_3\ldots\sin^2\theta_{D-1}}\times\mathrm{cosec}^2\theta_2\mathrm{cosec}\theta_1\frac{\partial f}{\partial\theta_2}$$

$$= -\cos\theta_1\cot\theta_2\frac{\partial f}{\partial\theta_1} - \sin\theta_1\frac{\partial f}{\partial\theta_2}$$

$$(B12)$$

Therefore

$$\left(x_2\frac{\partial}{\partial x_3} - x_3\frac{\partial}{\partial x_2}\right)^2 f$$

$$= \left(\cos\theta_1\cot\theta_2\frac{\partial}{\partial\theta_1} + \sin\theta_1\frac{\partial}{\partial\theta_2}\right)\left(\cos\theta_1\cot\theta_2\frac{\partial f}{\partial\theta_1} + \sin\theta_1\frac{\partial f}{\partial\theta_2}\right)$$

$$= -\sin\theta_1\cos\theta_1\cot^2\theta_2\frac{\partial f}{\partial\theta_1} + \cos^2\theta_1\cot^2\theta_2\frac{\partial^2 f}{\partial\theta_1^2}$$

$$+\cos^2\theta_1\cot\theta_2\frac{\partial f}{\partial\theta_2} + \sin\theta_1\cos\theta_1\cot\theta_2\frac{\partial^2 f}{\partial\theta_1\partial\theta_2}$$

$$- \sin \theta_1 \cos \theta_1 \operatorname{cosec}^2 \theta_2 \frac{\partial f}{\partial \theta_1}$$

$$+ \sin \theta_1 \cos \theta_1 \cot \theta_2 \frac{\partial^2 f}{\partial \theta_1 \partial \theta_2} + \sin^2 \theta_1 \frac{\partial^2 f}{\partial \theta_2^2} \qquad (B13)$$

Again

$$(x_1 p_3 - x_3 p_1)$$

$$= \sum_{j=0}^{D-1} \frac{x_1^2}{h_j^2} \frac{\partial}{\partial \theta_j} \left(\frac{x_3}{x_1} \right) \frac{\partial f}{\partial \theta_j}$$

$$= \frac{x_1^2}{h_1^2} \frac{\partial f}{\partial \theta_1} \cot \theta_2 \sec \theta_1 \tan \theta_1 + \frac{x_1^2}{h_2^2} \frac{\partial f}{\partial \theta_2} (-\operatorname{cosec}^2 \theta_2) \sec \theta_1$$

$$= \frac{r^2 \cos^2 \theta_1 \sin^2 \theta_2 \ldots \sin^2 \theta_{D-1}}{r^2 \sin^2 \theta_2 \ldots \sin^2 \theta_{D-1}} \cot \theta_2 \sec \theta_1 \tan \theta_1 \frac{\partial f}{\partial \theta_1}$$

$$- \frac{r^2 \cos^2 \theta_1 \sin^2 \theta_2 \sin^2 \theta_3 \ldots \sin^2 \theta_{D-1}}{r^2 \sin^2 \theta_3 \ldots \sin^2 \theta_{D-1}} \operatorname{cosec}^2 \theta_2 \sec \theta_1 \frac{\partial f}{\partial \theta_2}$$

$$= \sin \theta_1 \cot \theta_2 \frac{\partial f}{\partial \theta_1} - \cos \theta_1 \frac{\partial f}{\partial \theta_2}$$

$$(x_1 p_3 - x_3 p_1)^2 f$$

$$= \left(\sin \theta_1 \cot \theta_2 \frac{\partial}{\partial \theta_1} - \cos \theta_1 \frac{\partial}{\partial \theta_2} \right) \left(\sin \theta_1 \cot \theta_2 \frac{\partial}{\partial \theta_1} - \cos \theta_1 \frac{\partial}{\partial \theta_2} \right) f$$

$$= \sin\theta_1 \cos \theta_1 \cot^2 \theta_2 \frac{\partial f}{\partial \theta_1} + \sin^2 \theta_1 \cot^2 \theta_2 \frac{\partial^2 f}{\partial \theta_1^2}$$

$$+ \sin \theta_1 \cos \theta_1 \operatorname{cosec}^2 \theta_2 \frac{\partial f}{\partial \theta_1} - \cos \theta_1 \sin \theta_1 \cot \theta_2 \frac{\partial^2 f}{\partial \theta_1 \partial \theta_2}$$

$$+ \sin^2 \theta_1 \cot \theta_2 \frac{\partial f}{\partial \theta_2} - \sin \theta_1 \cos \theta_1 \cot \theta_2 \frac{\partial^2 f}{\partial \theta_1 \partial \theta_2} + \cos^2 \theta_1 \frac{\partial^2 f}{\partial \theta_2^2}$$
$$(B14)$$

Adding $(B13)$ and $(B14)$ we get

$$L_{13}^2 f + L_{23}^2 f$$

$$= -\left(\cot^2\theta_2\frac{\partial^2 f}{\partial\theta_1^2} + \cot\theta_2\frac{\partial f}{\partial\theta_2} + \frac{\partial^2 f}{\partial\theta_2^2}\right)$$

where we have used $(B10)$ and $(B11)$.

Hence from $(B9)$

$$L_2^2 f = -\left(\frac{\partial^2 f}{\partial\theta_1^2} + \cot^2\theta_2\frac{\partial^2 f}{\partial\theta_1^2} + \cot\theta_2\frac{\partial f}{\partial\theta_2} + \frac{\partial^2 f}{\partial\theta_2^2}\right)$$

$$= -\left(\operatorname{cosec}^2\theta_2\frac{\partial^2 f}{\partial\theta_1^2} + \cot\theta_2\frac{\partial f}{\partial\theta_2} + \frac{\partial^2 f}{\partial\theta_2^2}\right)$$

$$= -\left[\frac{1}{\sin^2\theta_2}\frac{\partial^2 f}{\partial\theta_1^2} + \frac{1}{\sin\theta_2}\left(\cos\theta_2\frac{\partial f}{\partial\theta_2} + \sin\theta_2\frac{\partial^2 f}{\partial\theta_2^2}\right)\right]$$

$$= -\left[\frac{1}{\sin^2\theta_2}\frac{\partial^2 f}{\partial\theta_1^2} + \frac{1}{\sin\theta_2}\frac{\partial}{\partial\theta_2}\left(\sin\theta_2\frac{\partial f}{\partial\theta_2}\right)\right]$$

$$= -\left[-\frac{L_1^2 f}{\sin^2\theta_2} + \frac{1}{\sin\theta_2}\frac{\partial}{\partial\theta_2}\left(\sin\theta_2\frac{\partial f}{\partial\theta_2}\right)\right]$$

$$(B15)$$

In general

$$L_k^2 = -\left[\frac{1}{\sin^{k-1}\theta_k}\frac{\partial}{\partial\theta_k}\sin^{k-1}\theta_k\frac{\partial}{\partial\theta_k} - \frac{L_{k-1}^2}{\sin^2\theta_k}\right] \qquad (B16)$$

Index